AF361924

Urban Heat Island
Mitigation Technologies

Urban Heat Island Mitigation Technologies

Editor

Rohinton Emmanuel

MDPI • Basel • Beijing • Wuhan • Barcelona • Belgrade • Manchester • Tokyo • Cluj • Tianjin

Editor
Rohinton Emmanuel
Glasgow Caledonian University
UK

Editorial Office
MDPI
St. Alban-Anlage 66
4052 Basel, Switzerland

This is a reprint of articles from the Special Issue published online in the open access journal *Energies* (ISSN 1996-1073) (available at: hhttps://www.mdpi.com/journal/energies/special_issues/UHIM).

For citation purposes, cite each article independently as indicated on the article page online and as indicated below:

LastName, A.A.; LastName, B.B.; LastName, C.C. Article Title. *Journal Name* **Year**, *Volume Number*, Page Range.

ISBN 978-3-0365-0722-4 (Hbk)
ISBN 978-3-0365-0723-1 (PDF)

Contents

About the Editor

Rohinton Emmanuel is a Professor of Sustainable Design and Construction at the School of Computing, Engineering and Built Environment and the Director of the Research Centre for Built Environment Asset Management (BEAM; https://www.gcu.ac.uk/assetmanagement/). He pioneered the inquiry into urban heat island studies in warm regions and has taught and consulted on climate and environment-sensitive design, building and urban sustainability and its assessment, building energy efficiency, thermal comfort, and carbon in the built environment. Rohinton was the secretary of the largest group of urban climate researchers, the International Association for Urban Climate (2010–2013), was a member of the Expert Team on Urban and Building Climatology (ET 4.4) of the World Meteorological Organization (WMO), as well as a member of the CIB Working Group (W108) on Buildings and Climate Change. He has authored over 150 research publications, including An Urban Approach to Climate Sensitive Design (E&FN Spon Press, 2005), Carbon Management in the Built Environment (Routledge, 2012), Critical Concepts in Built Environment: Sustainable Buildings (Routledge, 2014) and Urban Climate Challenges in the Tropics (Imperial College Press, 2016). In terms of urban climate education and training, Rohinton is currently the Coordinator of an Erasmus Mundus Joint Master's Degree Programme on urban climate and sustainability (MUrCS; www.murcs.eu).

Preface to "Urban Heat Island Mitigation Technologies"

Although the effects of urban form on local climate have been studied extensively in the last 50 years, deliberate design interventions to harness theoretical knowledge on climate-sensitive design on large (urban and neighbourhood) scales as well as specific policy options for mitigation remain few and far between. This is especially the case in the warm, humid regions of the world, where much of the unintended climate consequences of urbanization are felt and interventions to mitigate them are exceedingly rare.

This Special Issue of Energies highlights recent practices and policy developments in the area of urban heat island mitigation. Given the traction that climate change mitigation and adaptation currently enjoys in the public imagination and the role of cities in such mitigation, now is the opportune moment to highlight what works and why and to explore the policy and practical implications of both urban design and urban planning.

We highlight issues around land use/land cover and their optimization for urban heat islands. The links between heatwaves resulting from our warming climate and the problem of urban overheating is also discussed in this Special Issue. Furthermore, the role of nature-based solutions (especially green infrastructure) in enhancing outdoor comfort in a Mediterranean context is presented. A case study on microclimate design process in a developing country is described for its potential relevance to architectural and urban design for enhanced climate sensitivity. Finally, broader technical solutions such as radiative cooling are explored to reduce the need for cooling.

Rohinton Emmanuel
Editor

Article

Understanding the Role of Optimized Land Use/Land Cover Components in Mitigating Summertime Intra-Surface Urban Heat Island Effect: A Study on Downtown Shanghai, China

Yan-jun Guo, Jie-jie Han, Xi Zhao, Xiao-yan Dai * and Hao Zhang *

Laboratory for Applied Earth Observation and Spatial Analysis (LAEOSA), Department of Environmental Science and Engineering, Jiangwan campus, Fudan University, Shanghai 200438, China; 17210740049@fudan.edu.cn (Y.-j.G.); 18210740005@fudan.edu.cn (J.-j.H.); 19210740064@fudan.edu.cn (X.Z.)
* Correspondence: xiaoyandai@fudan.edu.cn (X.-y.D.); zhokzhok@163.com (H.Z.)

Received: 5 March 2020; Accepted: 24 March 2020; Published: 3 April 2020

Abstract: In this study, 167 land parcels of downtown Shanghai, China, were used to investigate the relationship between parcel-level land use/land cover (LULC) components and associated summertime intra-surface urban heat island (SUHI) effect, and further analyze the potential of mitigating summertime intra-SUHI effect through the optimized LULC components, by integrating a thermal sharpening method combining the Landsat-8 thermal band 10 data and high-resolution Quickbird image, statistical analysis, and nonlinear programming with constraints. The results show the remarkable variations in intra-surface urban heat island (SUHI) effect, which was measured with the mean parcel-level blackbody sensible heat flux in kW per ha (Mean_pc_BBF). Through measuring the relative importance of each specific predictor in terms of their contributions to changing Mean_pc_BBF, the influence of parcel-level LULC components on excess surface flux of heat energy to the atmosphere was estimated using the partial least square regression (PLSR) model. Analysis of the present and optimized parcel-level LULC components and their contribution to the associated Mean_pc_BBF were comparable between land parcels with varying sizes. Furthermore, focusing on the gap between the present and ideally optimized area proportions of parcel-level LULC components towards minimizing the Mean_pc_BBF, the uncertainties arising from the datasets and methods, as well as the implications for sustainable land development and mitigating the UHI effect were discussed.

Keywords: surface urban heat island effect; land use/land cover; partial least square regression; nonlinear programming; Shanghai; China

---

1. Introduction

It is projected that global urban dwellers will increase from the present 55% to 68% in 2050 [1]. The ongoing urbanization presents numerous challenges for the urban environment and human well-being [2,3], as evidenced by the dramatic land use/land cover (LULC) change for intensive human settlements inevitably altered hydrological circulation, climate regulation, soil conservation, and biodiversity preservation [4–11]. One of the environmental consequences of LULC change is the urban heat island (UHI) effect. Associated with human activities and climate change [12–19], the UHI effect significantly affects emissions of air pollutants and greenhouse gases [20], meteorological disasters, and health risks [21,22].

Since the 1830s, the meteorological station recorded air temperature (AT) has been widely used to monitor the UHI effect at canopy height, due to the advantage of long time series data for historical climate analysis. Usually, the weather stations were sparsely and unevenly distributed within and around the cities. They are not fully representative to reveal the UHI effect of differing local climate zones [23], especially in the circumstance that urban encroachment in the suburban/rural area caused the biased AT recorded at local weather stations, which was known as the so-called "city-entering" phenomenon of stations [24]. The land surface temperature (LST) is used as an alternative to identify the surface urban heat island (SUHI) effect [25]. In contrast to local weather stations with low spatial coverage, the satellite-borne thermal sensors with a wide instantaneous field of view (IFOV) can capture the thermal features of land surface quickly and provide a cost-effective solution. Since the 1970s, the satellite-borne thermal remote sensing has been a promising approach for monitoring city-level and regional SUHI effect worldwide. So far, there has been well-documented literature on the application of satellite-borne thermal sensors for multiple-scale SUHI studies, using thermal infrared bands acquired from the low-resolution sensors (~km) such as Geostationary Operational Environmental Satellite (GOES), Advanced Very High-Resolution Radiometer (AVHRR) and Moderate-resolution Imaging Spectroradiometer (MODIS) and mediate resolution sensors (60–120 m) such as Landsat-5/Theme Mapper (TM), Landsat-7/Enhanced Theme Mapper Plus (ETM+), and Landsat-8/Thermal Infrared Sensor (TIRS), as well as Advanced Spaceborne Thermal Emission and Reflection Radiometer (ASTER) [26–28].

However, on the applicability of satellite thermal remote sensing for SUHI studies, two questions should be addressed (1) how to transfer the spatially explicit information of intra-SUHI effect to the decision-making of land development and land parcel design? and (2) how to link the LST-indicated SUHI effect with the well-known canopy UHI effect? On the first question, when we turned our researching of interest into the urban area, it was found that these low and mediate resolution thermal bands were not suitable for monitoring the SUHI effect in urban settings with diverse LULC components, various urban morphology, and fine-scale land parcel design attributes [29–31]. In practice, the Advanced Thermal and Land Applications Sensor (ATLAS) operated by the American National Aeronautics and Space Administration (NASA) can capture the fine-scale thermal property of urban LULC components. Unfortunately, such aerial high-resolution thermal images (~10 m) are costly and very lacking for most of the SUHI studies. On the other hand, several trial studies were performed to thermally sharpen the coarse MODIS or GOES data to simulate the thermal signals of Landsat TM/ETM+ and ASTER in the urban area, but the LST products were still too coarse to indicate the SUHI effect. Alternatively, the combined usage of the high-resolution satellite/aerial images and mediate resolution satellite TIR bands may provide a practical way of mapping the urban thermal environment [31]. On the second question, the direct linkage between the LST-indicated intra-SUHI effect with the canopy UHI effect may be subject to many uncertainties. As an alternative, the LST can be used to estimate the excess surface flux of heat energy to the atmosphere, which was closely related with the surface warming [32], and thus, can help illustrate the relationship between SUHI effect and canopy UHI effect.

Nowadays, urban areas, which are subject to local warming effect or the combined local and global warming effects, are the fundamental units for climate change adaptation and mitigation [33]. Thus, focusing on the relationship between LULC components and artificial modification of urban climate, how cities cope with the UHI effect optimizing land use patterns towards sustaining regional ecosystems, will significantly affect UHI mitigation and adaption to climate change [34–40].

Taking downtown Shanghai, for example, which suffers from summertime extreme heat events, as a case study, this study aims to investigate the relationship between parcel-level LULC components and summertime intra-SUHI effect, and further find the solution to cooling the city and mitigating summertime intra-SUHI effect through the optimized LULC components. We expect the findings of this study will be beneficial for decision-makers who are engaged in mitigating the UHI effect and adapting to climate change.

2. Location of the Study Area

Downtown Shanghai, the economic center of China, is located between latitudes 31°32′ N–31°27′ N and longitudes 120°52′ E–121°45′ E. This region has a northern subtropical monsoon climate. The average annual temperature approximates 15 °C, with temperatures averaging 28 °C in the summer and 4 °C in the winter. The average annual precipitation ranges between 1000 and 1200 mm, with approximately 60% of the rainfall being received during spring and autumn. It covers an area of 4000 km², housing approximately 84.03% of a total of 24.18 million permanent residents within the boundary of greater Shanghai [41]. In this study, concerning previous studies of urban function zones (UFZs) in China [42–45], four typical UFZs of downtown Shanghai were selected (see Figure 1 and Table 1).

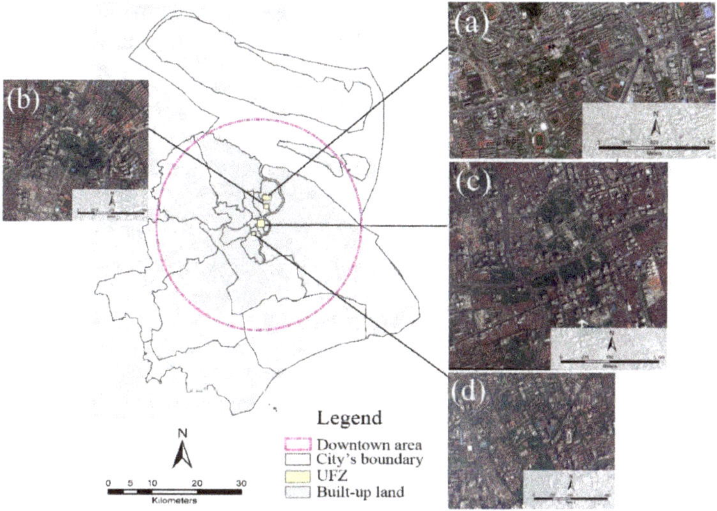

Figure 1. Location of the study area. Note: The labels (**a**), (**b**), (**c**), and (**d**) denote the four urban function zones (UFZs) known as Wujiaochang, PeacePark, UrbanCore, and Xujiahui, respectively.

Table 1. Description of four UFZs.

UFZ	Area (km²)	Description
Wujiaochang	7.09	One of the subcenters of downtown Shanghai with a cluster featuring a commercial center, colleges and universities, and a high-tech park for innovative firms.
PeacePark	2.00	Featured with a cluster of high education, innovative enterprises, Peace park, and recreational landscaping.
UrbanCore	5.97	The heart of downtown Shanghai, with a cluster of municipal administrative services, banking, headquarter economy, commercial center, and historical and cultural resorts.
Xujiahui	2.58	One of the sub-centers of downtown Shanghai, featuring a cluster of a commercial center, historical and cultural resorts, high education, advanced medical care, and innovative enterprises.

3. Materials and Methods

3.1. Materials

The data sources include the Landsat-8 images, high-resolution Quickbird image, and the auxiliary datasets. Due to the influence of cloud contamination, the available satellite images were limited.

In this study, two cloud-free Landsat-8 satellite images (path/row:118/38) acquired on two summertime dates (August 13, 2013, and August 3, 2015), were used for retrieval of LST. A high-resolution Quickbird image (dated April 1, 2014) covering the four UFZs was used for the classification of LULC components. Together with the free-accessible Google Earth and Baidu Map, the commercial digitalized urban thematic maps of downtown Shanghai [46] were used as the auxiliary datasets.

3.2. Methods

To better illustrate our research aims and related arrangement of the sections, an overall technical flowchart containing the basic and key procedures employed for this study were shown in Figure 2.

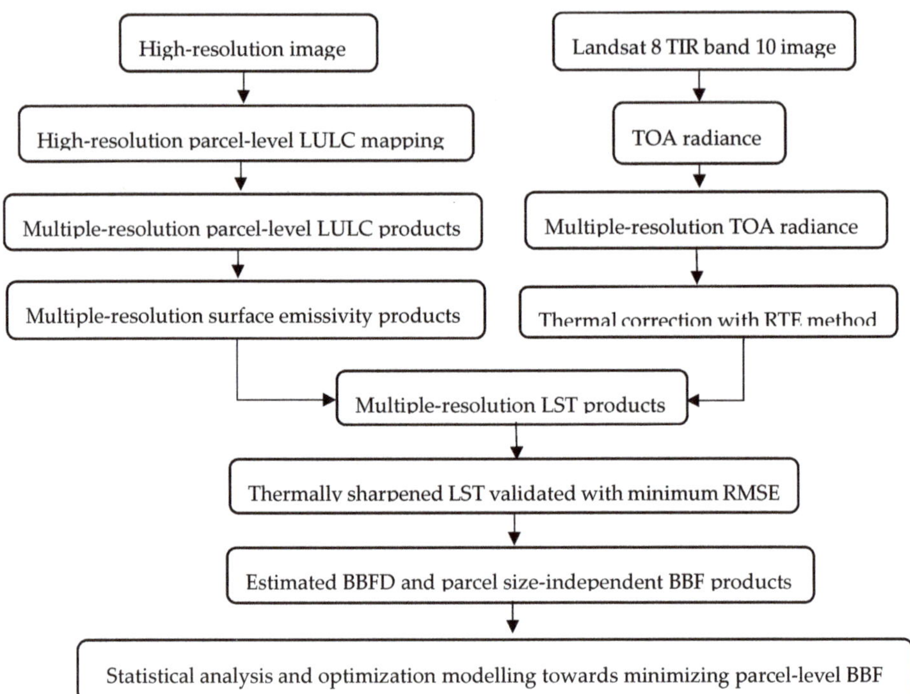

Figure 2. Flowchart of this study.

3.2.1. Classification of LULC Components and Delimitation of Land Parcels

By using a land surface classification system with nine LULC types in downtown Shanghai [31], the classification of LULC components was performed with an object-oriented classification (OOC) method. This OOC method could make full use of contextual data—such as spectral information, texture, spatial neighborhood properties, and fractal dimensions—from very high-resolution images and delineate the objects of interest [47] with much higher accuracy than traditional per-pixel classification methods [48–50]. The overall accuracy is approximately 80.03%. The post-classification product was further manually corrected by overlapping it with Google Earth and Baidu Map layers and then validated in a field survey, with an overall correction accuracy of 91.1%. Subsequently, based on our prior knowledge of land use and the developmental intensity of the UFZs, 167 regular polygons were drawn to delimit the land parcels along the directional and geometric features of traffic roads, bounding walls, and urban creeks enclosing the clustered land units.

Herein, we want to emphasize that the classification of LULC components was used for the estimation of land surface emissivity. Given the mixing-pixel effect of the land surface showing the same/similar LST, and many LULC types may complicate the analysis between LULC components and the associated thermal effect, for this study, the nine LULC classes were aggregated into four classes: building (architecture with vertical walls and roofs, e.g., house and warehouse), paved surface (e.g., asphalt road, square, playground), waterbody (e.g., creek and pond), and vegetation (e.g., tree, shrub, and lawn) (see Figure 3).

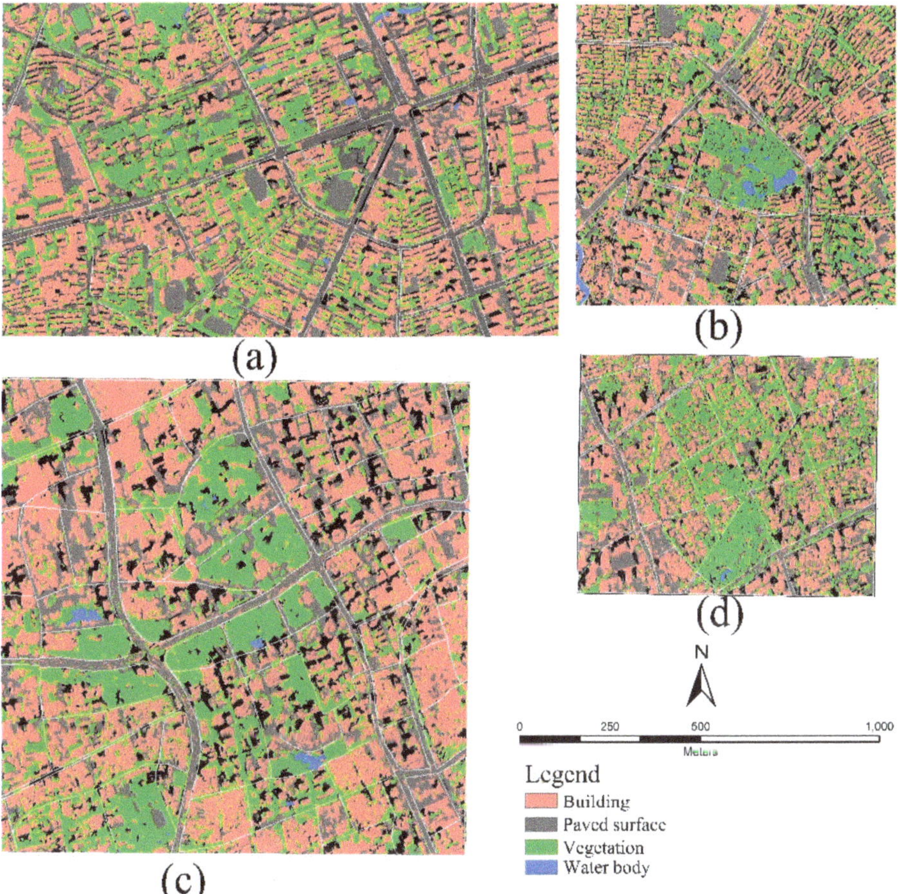

Figure 3. Parcel-level land use/land cover (LULC) components at four UFZs. Note: The labels (**a**), (**b**), (**c**), and (**d**) denote the four UFZs known as Wujiaochang, PeacePark, UrbanCore, and Xujiahui, respectively. The white polygons denote the land parcels' boundaries. The black spots denote the pixels shaded by the buildings and were removed to avoid the biased result of thermal analysis.

3.2.2. Retrieval and Validation of High-Resolution Thermally Sharpened LST

In this study, retrieval and validation of high-resolution thermally sharpened LST include the following steps. Firstly, for Landsat 8 TIR bands, given band 11 is subject to higher uncertainty of telescope stray light disturbance [51,52], band 10 was used to generate the top of the atmosphere (TOA) radiance as follows,

$$L_{sensor,\ \lambda} = gain \times DN + offset, \tag{1}$$

where $L_{sensor,\lambda}$ is the at-sensor radiance of thermal band pixels in W/(m^2 ster μm), the gain is the slope of the radiance/DN conversion function, and offset the slope of the radiance/DN conversion, respectively [53].

Secondly, a spatial interpolation-based method known as co-Kriging interpolation was employed to generate the high-resolution thermally sharpened LST, via a combination of the high-resolution land surface products and Landsat-8 TIR band 10 data [31]. Given the bias arising from the different resolutions between the high-resolution LULC components product and coarse TOA radiance data, both datasets must be scaled to the same resolution. Therefore, the high-resolution LULC components product were resampled with multiple resolutions (1–9 m) and set as the base maps to overlap and delimit the TOA radiance layer. We assumed that same/similar surfaces would have the same/similar radiance values since there is no available reference data for this study. Then, for each scene of the TOA radiance layer, we tried many times from hundreds of points per km^2 to 6000 points per km^2 to compare the co-Kriging interpolation results, and we found the remarkable decrease in the pairwise bias between the raw and simulated points in response to the increasing point number. It was found that the threshold of 3000 points was reasonable as the pairwise bias was inclined to get flat and changed little when the point number was greater than 3000 points [31]. Therefore, 3000 spatially random points per km^2 with the same interval as the multiple resolutions of the LULC component products were generated to extract the point-based radiance values falling within the polygon of each same/similar LULC category. The extracted points with radiance values were used to rebuild the TOA radiance layer, using the co-Kriging interpolation method. Besides, based on the multiple-resolution LULC layers, a surface emissivity correction for the same/similar LULC components was performed according to empirical studies [54,55] and laboratory testing [56]. Subsequently, based on a generalized single-channel method known as the range transfer equation (RTE) for retrieving LST [57], the emissivity corrected LST using was computed as follows,

$$L_{sensor,\lambda} = \left[\varepsilon_\lambda B_\lambda(T_s) + (1 - \varepsilon_\lambda) L^{\downarrow}_{atm,\lambda} \right] \times \tau_\lambda + L^{\uparrow}_{atm,\lambda} \tag{2}$$

$$B_\lambda(T_S) = \frac{L_{sensor,\,\lambda} - L^{\uparrow}_{atm,\,\lambda} - \tau_\lambda \times (1 - \varepsilon_\lambda) L^{\downarrow}_{atm,\,\lambda}}{\tau_\lambda \varepsilon_\lambda} \tag{3}$$

$$Ts = \frac{k_2}{\ln\left(1 + \frac{k_1}{B_\lambda(T_S)}\right)} \tag{4}$$

where $B_\lambda(T_S)$ is the black body radiance given by Plank's law, T_S the black body LST in Kelvin (K), ε_λ the corrected emissivity of the specific land surface (details see reference [31]). $L^{\downarrow}_{atm,\,\lambda}$ is the downwelling atmospheric radiance, $L^{\uparrow}_{atm,\,\lambda}$ the upwelling atmospheric radiance, τ_λ the total atmospheric transmissivity between the surface and sensor, all of which were retrieved from a web-based Atmospheric Correction Tool [58]. k_2 and k_1 are band 10 thermal conversion constants included in the metadata file of Landsat 8 data [53].

Finally, validation for the sharpened LST products was performed by comparing the pixel-based root-mean-square error (RMSE) between the original 30 m LST products and the resampled 30 m products from the sharpened products (Appendix A Table A1). The sharpened LST products at 1 m resolution were employed for further analysis due to their superior visual quality and minimal RMSE.

3.2.3. Estimation of Parcel-Level Sensible Heat Flux

Based on the high-resolution sharpened LST products, the parcel-level blackbody sensible heat flux (BBF) was estimated follows [32]:

$$\Phi_{BBFD} = \int_{10.60\,\mu m}^{11.19\,\mu m} \frac{C_1}{\pi \lambda^5 \left[\exp\left(\frac{C_2}{\lambda T}\right) \right]} d\lambda \tag{5}$$

$$BBF_i = \Phi_{BBFD} \times A_i \tag{6}$$

where Φ_{BBFD} is the pixel-based BBFD (W·m^{-2}), $C_1 = 3.7404 \times 10^8$ (W·μ^4·m^{-2}), $C_2 = 14387$, λ is TOA radiance, and T is LST in Kelvin (K), A_i is area of the ith land parcel (m^2).

However, due to the uneven size of the land parcels, there exists a remarkable variation of parcel-level BBF. Therefore, regardless of the size effect of land parcels, the parcel-level BBFs were converted to per ha BBF (pc_BBF), namely a cumulative number of the BBFD multiplying by the pixels within an idea parcel sized 1 ha, to make them comparable.

3.2.4. Statistical Analysis and Estimating the Optimized Lulc Components for Minimizing Mean_pc_BBF

As a regular step, the exploring data analysis procedure, including descriptive statistics, normality test, Box–Cox transformation for skewed data (if there were), and Pearson's product-moment correlation, was performed. In this study, Pearson's product-moment correlation analysis helped quantify the assumed relationship between the dependent and independent variables. However, the result of correlation analysis indicated the multicollinearity between the independent variables (for details see Section 4.1). Thus, the partial least square regression (PLSR) model that was developed for fixing the problem of the multilinearity was employed. Further, to avoid the over-fitting problem and determine a reasonable model with the appropriate number of components that has good predictive ability, the leave-one-out (LOO) method was used for cross-validation, by selecting the model with the highest average predicted R^2 and the lowest average prediction sum of squares (PRESS). The validated PLSR model was written as follows,

$$\text{Mean\_pc\_BBF} = \alpha_1 + \beta_1 \cdot X_1 + \beta_2 \cdot X_2 + \beta_3 \cdot X_3 + \beta_{23} \cdot X_2 \cdot X_3 + \beta_4 \cdot X_4 + \varepsilon, \tag{7}$$

where Mean_pc_BBF is the averaged pc_BBF of two summertime days since the overall parcel-level BBFD products on two dates were highly similar to each other (for details, see Appendix A Figure A1), so were the pc_BBF; α_1 is the intercept/constant item, $\beta_1 \sim \beta_4$ are the partial coefficients; $X_1 \sim X_4$ are the Box–Cox transformed area proportions of the paved surface, waterbody, vegetation, and building within each land parcel, respectively (for details see Appendix A Table A2). ε is the error term.

Subsequently, based on the validated PLSR model, nonlinear programming with constraints towards minimizing Mean_pc_BBF through the optimized LULC components was written as follows,

$$\text{Min} \quad \alpha_1 + \beta_1 \cdot X_1 + \beta_2 \cdot X_2 + \beta_3 \cdot X_3 + \beta_{23} \cdot X_2 \cdot X_3 + \beta_4 \cdot X_4 \tag{8}$$

$$\text{s.t. } X\prime 1 + X\prime 2 + X\prime 3 + X\prime 4 \leq 100 \tag{9}$$

$$s.t. \begin{cases} X\prime 1 + X\prime 2 + X\prime 3 + X\prime 4 \leq 100 & (9) \\ \alpha 0 + \beta\prime 1 \cdot X3 + \beta\prime 2 \cdot X3^2 + \beta\prime 3 \cdot X3^3 \geq \text{BBF\_Threshold} & (10) \end{cases} \tag{10}$$

where the definitions of α_1, $\beta_1 \sim \beta_4$, $X_1 \sim X_4$ are the same as Equation (7). $X'_1 \sim X'_4$ are the raw area proportions of the paved surface, waterbody, vegetation, and building within each land parcel. α_0 is the constant item of the cubic regression model, $\beta'_1 \sim \beta'_3$ are the coefficients. BBF_Threshold is the threshold value of pc_BBF in response to the changing area proportion of vegetation.

The statistical processes employed in this study were performed with R 3.6.2 [59] and the pls library for PLSR [60]. The Rsolnp library [61] was used to estimate the optimized LULC components for minimizing Mean_pc_BBF.

4. Results

4.1. Synoptic Analysis of Parcel-level LULC Components and Mean_pc_BBF

Overall, Figures 3 and 4 show the highly similar patterns of parcel-level LULC components and associated intra-SUHI effect measured with Mean_pc_BBF at four UFZs. As shown in Figure 3, there were significant variations of parcel-level LULC components measured by their area proportions. The buildings occupied the highest area proportion of the land parcels (averaged 47.72%), followed by the paved surfaces (averaged 24.59%) and vegetation (averaged 23.98%), whereas waterbody occupied the lowest area proportion (averaged 1.25%). In response to the fragmented vegetation and waterbodies within the total landscape, Figure 4 shows the overall unevenly distributed pattern of Mean_pc_BBF. As shown, the clusters of pixels with a paved surface and dense buildings exhibited higher BBFD than the spatially sparse pixels with waterbodies and vegetation. By contrast, the aggregated pixels with dominant vegetation and waterbodies, such as the university campus, big parks, boulevards, and a few well-vegetated residential communities, exhibited significantly lower Mean_pc_BBF (65.24–72.00 kW, averaged 68.47 kW); whereas most of the parcels dominated with paved surfaces and dense buildings exhibited the mediate Mean_pc_BBF (68.10–72.38 kW, averaged 70.31 kW) to higher Mean_pc_BBF (68.50–74.56 kW, averaged 70.70 kW).

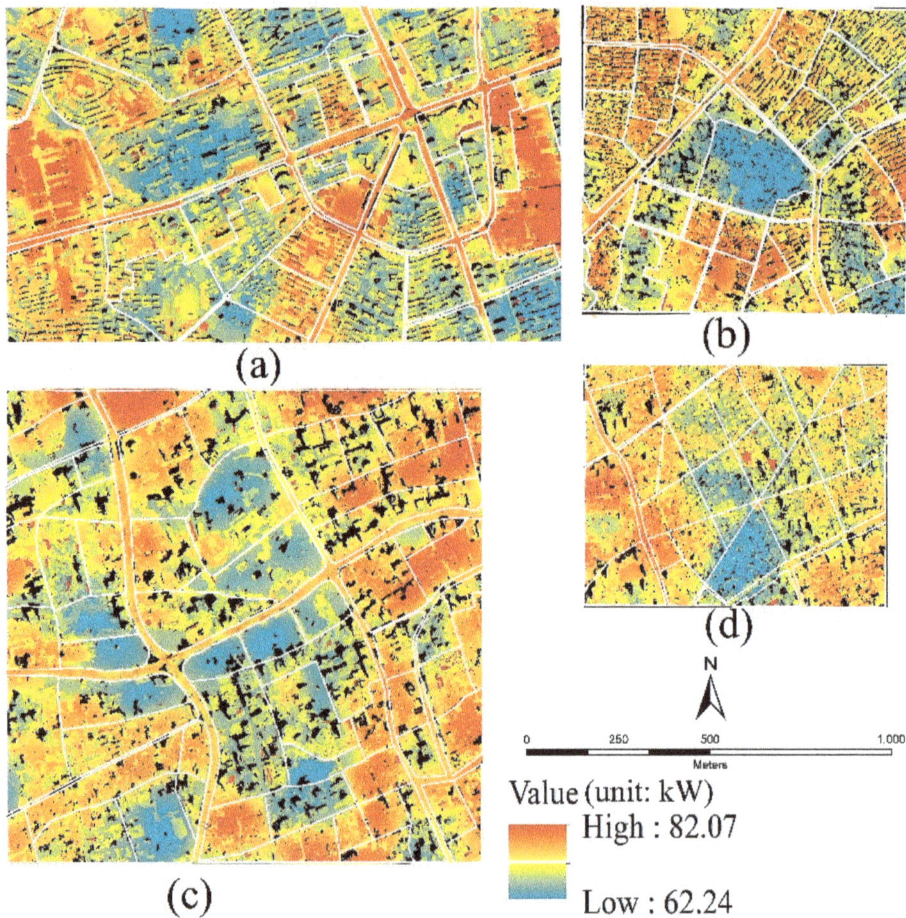

Figure 4. Parcel-level size-independent BBF (Mean_pc_BBF) at four UFZs. Note: The labels (**a**), (**b**), (**c**), and (**d**) denote the four UFZs known as Wujiaochang, PeacePark, UrbanCore, and Xujiahui, respectively. The definitions of white polygons and the black spots are the same as Figure 3.

4.2. Interpretation of Relationship between Parcel-Level LULC Components and Mean_pc_BBF

In the sense of qualitative statistics, Table 2 shows the statistically significant relationships between the Box–Cox transformed parcel-level LULC components and the intra-SUHI effect measured with Mean_pc_BBF, as evidenced with the correlation coefficients indicating the multicollinearity that may mislead our understanding of the relationships as mentioned above. However, it can be seen that except for the positive relationship between X2 and X3, there were the negative relationships between the other LULC components, indicating the overall exclusive competition of land use purpose for urban land development and associated influence on Mean_pc_BBF. Figure 5 visually addresses the bivariate relationships between fine-scale parcel-level LULC components (e.g., building and vegetation) and Mean_pc_BBF. The observed positive/negative bivariate relationships with remarkable variations in Mean_pc_BBF are wholly in agreement with the heterogeneity of parcel-level LULC components and associated Mean_pc_BBF shown in Figures 3 and 4. Based on Figure 5b, the BBF_Threshold in equation 10 was estimated.

Table 2. Coefficients of Pearson's product-moment correlations.

Pairwise	Correlation
X2 (Waterbody) vs. X1 (Paved surface)	−0.166*
X3 (Vegetation) vs. X1 (Paved surface)	−0.312**
X4 (Building) vs. X1 (Paved surface)	−0.194*
X3 (Vegetation) vs. X2 (Waterbody)	0.503**
X4 (Building) vs. X2 (Waterbody)	−0.387**
X4 (Building) vs. X3 (Vegetation)	−0.611**
X1 (Paved surface) vs. Mean_pc_BBF	0.620**
X2 (Waterbody) vs. Mean_pc_BBF	−0.435**
X3 (Vegetation) vs. Mean_pc_BBF	−0.470**
X4 (Building) vs. Mean_pc_BBF	0.639**

Note: In this table, all the independent variables were Box–Cox transformed. * and ** denote significance at 0.05 and 0.01 levels, respectively.

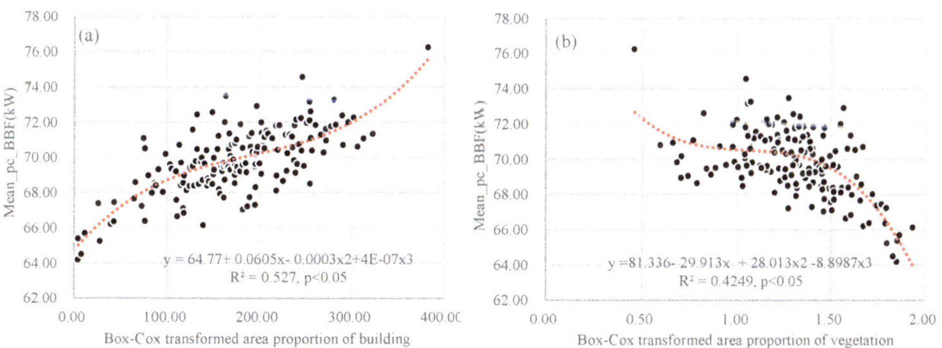

Figure 5. Non-linear regression models showing the statistical relationships between (**a**) Box–Cox transformed area proportion of building and Mean_pc_BBF; (**b**) Box–Cox transformed area proportion of vegetation and Mean_pc_BBF. Note: Given the high correlation between the Box–Cox transformed variables and Mean_pc_BBF (see Table 2) that causes the multicollinearity, only two of them were selected to draw the scatterplots.

Further, beyond addressing the bivariate relationships shown in Table 2 and Figure 5, Table 3 comprehensively shows the quantitative relationship between the Box–Cox transformed parcel-level LULC components and their roles in controlling Mean_pc_BBF, explaining 56.1% variation of the Mean_pc_BBF associated with independent variables. As shown, along with the waterbody, the interaction of waterbody and vegetation exhibited the statistically significant negative coefficients, indicating their capacity for suppressing Mean_pc_BBF when fixing the other variables. In contrast, paved surface, vegetation, and building exhibited statistically significant positive coefficients, indicating their capacity for increasing Mean_pc_BBF. However, herein, the unstandardized coefficients (Coefs) of the Box–Cox transformed variables result in puzzling understanding as they cannot be directly comparable for indicating each specific positive/negative relationship with Mean_pc_BBF.

Table 3. Coefficients of partial least square regression (PLSR) models.

Variable	Coef	S-Coef
Constant	64.699	0
X1: Paved surface	0.417	0.267
X2: Waterbody	−0.464	−0.291
X3: Vegetation	0.562	0.085
X4: Building	0.019	0.741
Waterbody × Vegetation	−0.32	−0.196
Summary statistics	F = 51.340, p < 0.05	
	Adjusted R^2 = 0.561	

Note: Coef and S-Coef denote unstandardized and standardized coefficients, respectively. In this table, all the independent variables were Box–Cox transformed.

Alternatively, by measuring the changing standard deviation of Mean_pc_BBF in response to each one standard deviation increase in a given Box–Cox transformed variable, the standardized coefficients (S-Coefs) can better indicate the relative strength of the relationship between the Box–Cox transformed variables and Mean_pc_BBF. As indicated, the waterbody was found to be the strongest negative determinant of Mean_pc_BBF when fixing the other variables, as with each one standard deviation increase in waterbody would approximately decrease 0.291 standard deviations of Mean_pc_BBF. The interaction of waterbody and vegetation was found to be the second strongest negative determinant of Mean_pc_BBF when fixing the other variables, as with each one standard deviation increase in the interaction of waterbody and vegetation would approximately decrease 0.196 standard deviations of Mean_pc_BBF. Besides, vegetation was found to be the smallest positive determinant of Mean_pc_BBF when fixing the other variables, as with each one standard deviation increases in vegetation would approximately increase 0.085 standard deviations of Mean_pc_BBF. In contrast, when fixing the other variables, paved surface and building were found to be the mediate and highest positive determinants of Mean_pc_BBF, as with the relatively remarkable increase in standard deviations of Mean_pc_BBF, which respond to each one standard deviation increase in paved surface and building, respectively. The use of the S-Coefs made sense for the results.

4.3. Comparison of the Present and Optimized Parcel-level LULC Components and the Associated Mean_pc_BBF

Table 4 shows the remarkable changes in summertime Mean_pc_BBFs associated with present and optimized parcel-level LULC components. As shown, except for the slight change of proportion of the other impervious surface in the present and optimized parcel-level LULC components, the changes in area proportions of the other LULC components were remarkable. As a whole, compared with present LULC components, the optimized parcel-level LULC components with the thresholding area proportions would cause an overall 51.80% decrease in summertime Mean_pc_BBFs.

On the other hand, if fixing the area portion of the other impervious surface constant, it is an excellent way to enhance the parcel-level area proportions of the waterbody and vegetation and simultaneously decrease area proportion of Building. However, due to the remarkable variation of parcel-level LULC components, there were only 7.79% of the land parcels (big parks and memorial cemetery), ultimately meeting the optimized thresholding area proportions of the parcel-level LULC components. In contrast, the land parcels with dominant impervious surface approximated 44.31% of total land parcels. Then, focusing on the gap between the present and ideally optimized area proportions of parcel-level LULC components, there would be higher uncertainties of minimizing the Mean_pc_BBF through adjusting urban land development and optimizing the parcel-level LULC components. How to seek practical solutions will be further emphasized in the discussion section.

Table 4. Changes in summertime Mean_pc_BBFs associated with present and optimized parcel-level LULC components.

	Area Proportion (%)		Mean_pc_BBF (kW)	
LULC	Present	Optimized	Present	Optimized
Other	24.59	24.68		
Waterbody	1.259	4.97	128.01	61.70
Vegetation	23.98	42.91		
Building	47.72	27.44		

5. Discussion

5.1. On the Data, Methods, and Findings of this Study

The occurrence of the intra-UHI effect depends on the complicated human-nature processes in the built environment. In addition to the macro-scale climate system, the micro-scale artificial modification of urban climate due to varying thermal properties of LULC is one of the key drivers of the intra-SUHI effect. Most of the datasets used in this study are publicly accessible, and statistical analysis methods and optimization algorithm adopted are well known among the researchers. Together with the spatial interpolation-based thermal sharpening method, which was proven to be capable to capture the varying thermal properties of fine-scale LULC components [31], the PLSR analysis quantitatively examined the relative importance of parcel-level LULC components in determining the Mean_pc_BBF. Furthermore, results of the present and optimized parcel-level LULC components and their contribution to the associated Mean_pc_BBF were comparable across the land parcels, regardless of their varying size.

In the existing studies for investigating SUHI effect via satellite thermal remote sensing, most of which only used the MODIS, Landsat TM/ETM+/8, and ASTER thermal band products [62,63], or additionally combined with high-resolution aerial photos/commercial satellite satellites for added interpretation [17,64,65], the results based on the unsharpened thermal band data were too coarse to examine the relationship between the observed LST and fine-scale LULC components. In contrast, our findings, though exhibiting the uncertainties of the Mean_pc_BBF between the present and ideally optimized area proportions of parcel-level LULC components, can provide more detailed information on the fine-scale intra-SUHI effect and optional choices for decision-making towards mitigating UHI effect. Thus, the application of the accessible public datasets and the methods in this study can be referenced to the case studies elsewhere.

5.2. Implications for Practical Solutions to Optimized Parcel-Level LULC Components towards Mitigating Intra-SUHI Effect

In the modern history of downtown Shanghai, this city was in the track of international trading and industrialization manipulated by the western colonialist authorities. This city's compact form had been criticized for deteriorating the natural watery landscape and lacking green space during its growth [66]. In the period of 'urban industrialization (1949–2000)', this city became China's biggest industrial center, and most of the land patches were encouraged for intensive development of industrial facilities, traffic roads, and residential quarters at the loss of the creeks and farmlands within the former urban-rural mosaics [67–69]. The resultant remnants of waterbodies and vegetation within urban areas accounted for tiny area proportions of downtown Shanghai, making this city a typically hot city during the summer.

Recently, with the growing awareness of the better environment, the local government has been dedicated to urban regeneration and improve this city's green infrastructure via land replacement for new greenspace. The consequent increase in greenspace from 8278 ha in 1998 to 127,332 ha in 2015 was rewarded with a remarkable decrease in previous UHI hotspots [70–72]. However, as addressed in Section 4, the overall competitive parcel-level LULC components embodied the heterogeneity of urban land development intention and associated SUHI effect. As evidence, the cumulative area proportions of waterbodies and vegetation are still less than the recommended minimum 30% baseline of the study area, which can better interpret the weakness of present parcel-level waterbodies and vegetation in modifying the SUHI effect. To achieve the goal of minimizing Mean_pc_BBF, it is essential to trade-off the conflicts rooting in the facets of urban land development intention, then further reduce and remove the significant mismatching in area proportions between the present and optimized parcel-level LULC components. To do so, regeneration of the land parcels with dominant buildings and hard-top pavements well as architectural renovation should be officially given the top priority. For example, practice in ecological resilience, such as the well-managed vertical planting and roof garden, has exemplified its remarkable cooling capacity in microclimatic UHI mitigation. In addition to increasing the green ratio, it is crucial to shape the openness of urban morphology and alleviate the blocking of wind corridors that reduces convective heat removal and transfer by the wind [73]. We recommend that future land parcel design should guide the moderate density clusters of middle- and high-rises with acceptable building distance (e.g., 30 m or more, rather than merely the sunshine spacing).

5.3. Limitation of this Study and Future Researching Tasks

There are several noticeable shortcomings of this study. Firstly, due to the 16-day revisiting interval of Landsat 8 satellite and cloud contamination, only two cloud-free images are available for this study. Thus, the estimated parcel-level BBF, which was generated from the thermally sharpened LST products, can only capture the instantaneous field of view rather than the continuous scenes of the study area. Secondly, until the present, except for a few five-year intervals of aerial photography performed in earlier years, there have no available airborne sensors like the ATLAS at 10-m for investigating the BBF in downtown Shanghai. Alongside with lacking the airborne sensed data, the absence of in-situ measured parcel-level BBF datasets made the cross-validation of the estimated parcel-level BBF impossible. Thirdly, the three-dimensional features of urban morphology and their influences on micro-climate and human thermal comfort were not considered in this study. Consequently, in the sense of data demand for near real-time, temporal continuity, spatial representativeness, the direct linkage between the retrieved parcel-level BBF, human thermal comfort, and energy requirement for cooling was insufficiently illustrated. To fill these emphasized knowledge gaps, in future research design, the in-situ measurement of micro-climatic parameters, high-resolution land cover types, the three-dimensional features of urban morphology, should be incorporated into the computational fluid dynamics (CFD) simulation platforms (e.g., ENVI-met and WindperfectDX) to cross-validate the satellite observation and in-situ measurements, and then be further refined to generate the tempo-spatially continuous datasets.

In short, we argue that focusing on these above-mentioned shortcomings, to find the good questions and attempt to comprehensively put the conceived ideas into practice, then we can get close to the truth and find the practical ways to mitigate the adverse UHI effect and further enhance the adaption capacity to climate change.

6. Conclusions

Selecting downtown Shanghai as a case, this study quantitatively examined the influences of parcel-level LULC components on summertime intra-SUHI effect and further analyzed the potential of mitigating summertime intra-SUHI effect through the optimized LULC components. The major findings were summarized as follows,

(1) Overall, there exists a statistically significant relationship between fine-scale parcel-level LULC components and the intra-SUHI effect measured with Mean_pc_BBF. The observed remarkable variations in Mean_pc_BBF can be attributed to the heterogeneity of parcel-level LULC components.

(2) Focusing on the influence of parcel-level LULC components on intra-SUHI effect, the relative importance of waterbodies and vegetation in terms of their contributions to decreasing associated Mean_pc_BBF urgently require increasing their area proportions. Theoretically, the optimized parcel-level LULC components would cause an estimated 51.798% decrease in summertime Mean_pc_BBFs. However, how to balance the conflicts between the present and ideally optimized parcel-level LULC components towards minimizing the Mean_pc_BBF should be carefully considered. It is still a long and formidable challenge for decision-making of sustainable land development and mitigating the UHI effect.

(3) In contrast to the relative coarse scales of region and city levels, our study demonstrated a practical approach linking fine-scale parcel-level LULC components and intra-SUHI effect, using the integration of satellite-based thermal sharpening, statistical analysis, and nonlinear programming with constraints. The methodology and findings presented beneficial insights for guiding sustainable urban land development towards enhancing the city's adaption capacity to climatic change for megacities like Shanghai.

Author Contributions: H.Z. conceived the central idea and designed the technical framework for this study. Y.-j.G., J.-j.H., X.Z., X.-y.D., and H.Z. commanded data processing and analysis. H.Z. wrote the manuscript. X.-y.D. and H.Z. reviewed and edited the draft manuscript. All authors have read and agreed to the published version of the manuscript.

Funding: This study is supported by the Original and Innovatory Research Program (2017–2019) of Fudan University.

Acknowledgments: The authors thank the Geospatial Data Cloud site, Computer Network Information Center, Chinese Academy of Sciences (http://www.gscloud.cn) for providing the free Landsat TM/OLI images. The authors are indebted to the R Foundation for Statistical Computing and Beijing Piesat Information Technology Co., Ltd for free usage of the PIE 6.0 Remote Sensing image processing system.

Conflicts of Interest: The authors declare no conflict of interest.

Appendix A

Table A1. Pixel-based root-mean-square error (RMSE) between the original 30 m and the resampled 30 m LST products (1–9 m).

	1m	3m	5m	7m	9m
Mean ± s.d.	2.51 ± 0.32^{a}	2.71 ± 0.33^{a}	$2.83 \pm 0.42^{a,b}$	$2.90 \pm 0.56^{a,b}$	3.20 ± 0.65^{b}

Note: The different labels denote significant difference at 0.05 level.

Table A2. Estimations of the lambda (λ) and formulas.

Box-Cox Transformed	Estimated Lambda (λ)	Formulas
X1	0.50	$X1 = \text{sqrt (Paved surface)}$
X2	0.11	$X2 = \dfrac{(\text{Water body})^{0.11}}{0.89}$
X3	0.00	$X3 = \log (\text{Vegetation})$
X4	1.34	$X2 = \dfrac{(\text{Building})^{0.11}}{0.89}$

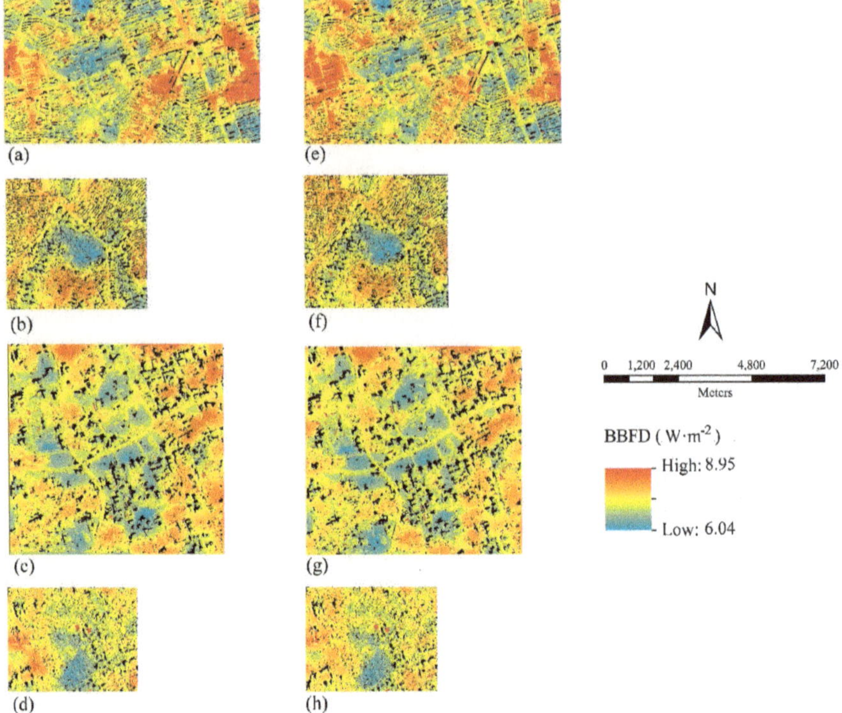

Figure A1. The UFZ-specific BBFD maps on two summertime dates. Note: (**a**), (**b**), (**c**), and (**d**) denote the UFZ-specific BBFD (dated on August 13, 2013) at Wujiaochang, PeacePark, UrbanCore, and Xujiahui, respectively. (**e–h**) denotes UFZ-specific BBFD (dated on August 3, 2015), respectively.

References

1. United Nations. 2018 Revision of World Urbanization Prospects. Available online: https://www.un.org/development/desa/publications/2018-revision-of-world-urbanization-prospects.html (accessed on 20 November 2019).
2. Luck, M.; Wu, J. A gradient analysis of urban landscape pattern: A case study from the Phoenix metropolitan region, Arizona, USA. *Landsc. Ecol.* **2002**, *17*, 327–339. [CrossRef]
3. Li, J.; Song, C.; Cao, L.; Zhu, F.; Meng, X.; Wu, J. Impacts of landscape structure on surface urban heat islands: A case study of Shanghai, China. *Remote Sens. Environ.* **2011**, *115*, 3249–3263. [CrossRef]
4. Cai, Y.-B.; Li, H.-M.; Ye, X.-Y.; Zhang, H. Analyzing Three-Decadal Patterns of Land Use/Land Cover Change and Regional Ecosystem Services at the Landscape Level: Case Study of Two Coastal Metropolitan Regions, Eastern China. *Sustainability* **2016**, *8*, 773. [CrossRef]
5. Clerici, N.; Cote-Navarro, F.; Escobedo, F.J.; Rubiano, K.; Villegas, J.C. Spatio-temporal and cumulative effects of land use-land cover and climate change on two ecosystem services in the Colombian Andes. *Sci. Total Environ.* **2019**, *10*, 1181–1192. [CrossRef] [PubMed]
6. Dewan, A.M.; Yamaguchi, Y. Land use and land cover change in Greater Dhaka, Bangladesh: Using remote sensing to promote sustainable urbanization. *Appl. Geogr.* **2009**, *29*, 390–401. [CrossRef]
7. King, R.S.; Scoggins, M.; Porras, A. Stream biodiversity is disproportionately lost to urbanization when flow permanence declines: Evidence from southwestern North America. *Freshw. Sci.* **2016**, *35*, 340–352. [CrossRef]

8.   Nguyen, H.H.; Recknagel, F.; Meyer, W. Effects of projected urbanization and climate change on flow and nutrient loads of a Mediterranean catchment in South Australia. *Ecohydrol. Hydrobiol.* **2019**, *19*, 279–288. [CrossRef]

9.   Pickard, B.R.; Van Berkel, D.; Petrasova, A.; Meentemeyer, R.K. Forecasts of urbanization scenarios reveal trade-offs between landscape change and ecosystem services. *Landsc. Ecol.* **2016**, *32*, 617–634. [CrossRef]

10.   Mati, B.M.; Mutie, S.; Gadain, H.; Home, P.; Mtalo, F. Impacts of land-use/cover changes on the hydrology of the transboundary Mara River, Kenya/Tanzania. *Lakes Reserv. Res. Manag.* **2008**, *13*, 169–177. [CrossRef]

11.   Reyers, B.; O'Farrell, P.J.; Cowling, R.M.; Egoh, B.N.; Le Maitre, D.C.; Vlok, J.H.J. Ecosystem services, land-cover change, and stakeholders: Finding a sustainable foothold for a semiarid biodiversity hotspot. *Ecol. Soc.* **2009**, *14*, 38. [CrossRef]

12.   McMichael, A.J.; Woodruff, R.E.; Hales, S. Climate change and human health: Present and future risks. *Lancet* **2006**, *367*, 859–869. [CrossRef]

13.   Jenerette, G.D.; Harlan, S.L.; Buyantuev, A.; Stefanov, W.L.; Declet-Barreto, J.; Ruddell, B.L.; Myint, S.W.; Kaplan, S.; Li, X. Micro-scale urban surface temperatures are related to land-cover features and residential heat related health impacts in Phoenix, AZ USA. *Landsc. Ecol.* **2016**, *31*, 745–760. [CrossRef]

14.   Chen, Y.; Cai, Y.; Tong, C. Quantitative analysis of urban cold island effects on the evolution of green spaces in a coastal city: A case study of Fuzhou, China. *Environ. Monit. Assess.* **2019**, *191*, 121. [CrossRef] [PubMed]

15.   Conlon, K.; Monaghan, A.; Hayden, M.; Wilhelmi, O. Potential Impacts of Future Warming and Land Use Changes on Intra-Urban Heat Exposure in Houston, Texas. *PLoS ONE* **2016**, *11*. [CrossRef]

16.   Huang, Q.; Lu, Y. The Effect of Urban Heat Island on Climate Warming in the Yangtze River Delta Urban Agglomeration in China. *Int. J. Environ. Res. Public Health* **2015**, *12*, 8773–8789. [CrossRef]

17.   Morabito, M.; Crisci, A.; Georgiadis, T.; Orlandini, S.; Munafò, M.; Congedo, L.; Rota, P.; Zazzi, M. Urban Imperviousness Effects on Summer Surface Temperatures Nearby Residential Buildings in Different Urban Zones of Parma. *Remote Sens.* **2018**, *10*, 26. [CrossRef]

18.   Trlica, A.; Hutyra, L.R.; Schaaf, C.L.; Erb, A.; Wang, J.A. Albedo, Land Cover, and Daytime Surface Temperature Variation Across an Urbanized Landscape. *Earth Future* **2017**, *5*, 1084–1101. [CrossRef]

19.   Turner, B.L., II. Land system architecture for urban sustainability: New directions for land system science illustrated by application to the urban heat island problem. *J. Land Use Sci.* **2016**, *11*, 689–697. [CrossRef]

20.   Harlan, S.L.; Ruddell, D.M. Climate change and health in cities: Impacts of heat and air pollution and potential co-benefits from mitigation and adaptation. *Curr. Opin. Env. Sust.* **2011**, *3*, 126–134. [CrossRef]

21.   Li, D.; Bou-Zeid, E. Synergistic Interactions between Urban Heat Islands and Heat Waves: The Impact in Cities Is Larger than the Sum of Its Parts. *Meteorol. Clim.* **2013**, *52*, 2051–2064. [CrossRef]

22.   Tan, J.; Zheng, Y.; Tang, X.; Guo, C.; Li, L.; Song, G.; Zhen, X.; Yuan, D.; Kalkstein, A.J.; Li, F. The urban heat island and its impact on heat waves and human health in Shanghai. *Int. J. Biometeorol.* **2010**, *54*, 75–84. [CrossRef] [PubMed]

23.   Stewart, I.D.; Oke, T.R. Local Climate Zones for Urban Temperature Studies. *B Am. Meteorol. Soc.* **2012**, *93*, 1879–1900. [CrossRef]

24.   Yang, Y.J.; Wu, B.W.; Shi, C.E.; Zhang, J.H.; Li, Y.B.; Tang, W.A.; Wen, H.Y.; Zhang, H.Q.; Shi, T. Impacts of urbanization and station-relocation on surface air temperature series in Anhui Province, China. *Pure Appl. Geophys.* **2013**, *170*, 1969–1983. [CrossRef]

25.   Benas, N.; Chrysoulakis, N.; Cartalis, C. Trends of urban surface temperature and heat island characteristics in the Mediterranean. *Theor. Appl. Climatol.* **2017**, *130*, 807–816. [CrossRef]

26.   Weng, Q. Thermal infrared remote sensing for urban climate and environmental studies: Methods, applications, and trends. *ISPRS J. Photogramm. Remote Sens.* **2009**, *64*, 335–344. [CrossRef]

27.   Bonafoni, S.; Anniballe, R.; Gioli, B.; Toscano, P. Downscaling Landsat Land Surface Temperature over the urban area of Florence. *Eur. J. Remote Sens.* **2016**, *49*, 553–569. [CrossRef]

28.   Holderness, T.; Barr, S.; Dawson, R.; Hall, J. An evaluation of thermal Earth observation for characterizing urban heatwave event dynamics using the urban heat island intensity metric. *Int. J. Remote Sens.* **2013**, *34*, 864–884. [CrossRef]

29.   Streutker, D. Satellite-measured growth of the urban heat island of Houston, Texas. *Remote Sens. Environ.* **2003**, *85*, 282–289. [CrossRef]

30. Baldinelli, G.; Bonafoni, S.; Anniballe, R.; Presciutti, A.; Gioli, B.; Magliulo, V. Spaceborne detection of roof and impervious surface albedo: Potentialities and comparison with airborne thermography measurements. *Solar Energy* **2016**, *49*, 553–569. [CrossRef]

31. Zhang, H.; Jing, X.-M.; Chen, J.-Y.; Li, J.-J.; Schwegler, B. Characterizing Urban Fabric Properties and Their Thermal Effect Using QuickBird Image and Landsat 8 Thermal Infrared (TIR) Data: The Case of Downtown Shanghai, China. *Remote Sens.* **2016**, *8*, 541. [CrossRef]

32. Stone, B.; Norman, J.M. Land use planning and surface heat island formation: A parcel-based radiation flux approach. *Atmos. Environ.* **2006**, *40*, 3561–3573. [CrossRef]

33. Georgescu, M.; Morefield, P.E.; Bierwagen, B.G.; Weaver, C.P. Urban adaptation can roll back warming of emerging megapolitan regions. *Proc. Natl. Acad. Sci. USA* **2014**, *111*, 2909–2914. [CrossRef] [PubMed]

34. Liu, J.; Shao, Q.; Yan, X.; Fan, J.; Zhan, J.; Deng, X.; Kuang, W.; Huang, L. The climatic impacts of land use and land cover change compared among countries. *J. Geogr. Sci.* **2016**, *26*, 889–903. [CrossRef]

35. Peng, J.; Jia, J.; Liu, Y.; Li, H.; Wu, J. Seasonal contrast of the dominant factors for spatial distribution of land surface temperature in urban areas. *Remote Sens. Environ.* **2018**, *215*, 255–267. [CrossRef]

36. Li, F.; Liu, X.; Zhang, X.; Zhao, D.; Liu, H.; Zhou, C.; Wang, R. Urban ecological infrastructure: An integrated network for ecosystem services and sustainable urban systems. *J. Clean Prod.* **2017**, *163*, S12–S18. [CrossRef]

37. Dissanayake, D.; Morimoto, T.; Murayama, Y.; Ranagalage, M. Impact of Landscape Structure on the Variation of Land Surface Temperature in Sub-Saharan Region: A Case Study of Addis Ababa using Landsat Data (1986–2016). *Sustainability* **2019**, *11*, 2257. [CrossRef]

38. Daily, G. *Nature's Services: Societal Dependence on Natural Ecosystems*; Island Press: Washington, DC, USA, 1997.

39. Gill, S.; Handley, J.; Ennos, A.; Pauleit, S. Adapting Cities for Climate Change: The Role of the Green Infrastructure. *Built. Environ.* **2007**, *33*, 115–133. [CrossRef]

40. Polydoros, A.; Mavrakou, T.; Cartalis, C. Quantifying the Trends in Land Surface Temperature and Surface Urban Heat Island Intensity in Mediterranean Cities in View of Smart Urbanization. *Urban Sci.* **2018**, *2*, 16. [CrossRef]

41. Shanghai Municipal Statistics Bureau (SMSB); Survey Office of the National Bureau of Statistics in Shanghai (SONBS-SH). *Shanghai Statistical Yearbook-2018*; China Statistics Press: Beijing, China, 2018.

42. Feng, Y.; Du, S.; Myint, S.W.; Shu, M. Do Urban Functional Zones Affect Land Surface Temperature Differently? A Case Study of Beijing, China. *Remote Sens.* **2019**, *11*, 1802. [CrossRef]

43. Huang, X.; Wang, Y. Investigating the effects of 3D urban morphology on the surface urban heat island effect in urban functional zones by using high-resolution remote sensing data: A case study of Wuhan, Central China. *ISPRS J. Photogramm.* **2019**, *152*, 119–131. [CrossRef]

44. Song, J.; Lin, T.; Li, X.; Prishchepov, A.V. Mapping urban functional zones by integrating very high spatial resolution remote sensing imagery and points of interest: A case study of Xiamen, China. *Remote Sens.* **2018**, *10*, 1737. [CrossRef]

45. Zhang, X.; Du, S.; Wang, Q.; Zhou, W. Multiscale geoscene segmentation for extracting urban functional zones from VHR satellite images. *Remote Sens.* **2018**, *10*, 281. [CrossRef]

46. Beijing Digital Space Technology Co. Ltd. *The Standard GIS-Based Altas of Shanghai*; Beijing Digital Space Technology Co. Ltd.: Beijing, China, 2015.

47. Blaschke, T.; Strobl, J. What's wrong with pixels? Some recent developments interfacing remote sensing and GIS. *GIS–Z. Geoinform. Syst.* **2001**, *14*, 12–17.

48. Hamedianfar, A.; Shafri, H.Z.M. Development of fuzzy rule-based parameters for urban object-oriented classification using very high resolution imagery. *Geocarto Int.* **2014**, *29*, 268–292. [CrossRef]

49. Mathieu, R.; Freeman, C.; Aryal, J. Mapping private gardens in urban areas using object-oriented techniques and very high-resolution satellite imagery. *Landsc. Urban Plan.* **2007**, *81*, 179–192. [CrossRef]

50. Su, W.; Zhang, C.; Yang, J.; Wu, H.; Deng, L.; Ou, W.; Yue, A.; Chen, M. Analysis of wavelet packet and statistical textures for object-oriented classification of forest-agriculture ecotones using SPOT 5 imagery. *Int. J. Remote Sens.* **2012**, *33*, 3557–3579. [CrossRef]

51. Malaret, E.; Bartolucci, L.A.; Lozano, D.F.; Anuta, P.E.; McGillem, C.D. Landsat-4 and Landsat-5 Thematic Mapper data quality analysis. *Photogramm. Eng. Rem. S.* **1985**, *51*, 1407–1416.

52. Barsi, J.; Schott, J.; Hook, S.; Raqueno, N.; Markham, B.; Radocinski, R. Landsat-8 Thermal Infrared Sensor (TIRS) Vicarious Radiometric Calibration. *Remote Sens.* **2014**, *6*, 11607–11626. [CrossRef]

53. U.S. Department of the Interior (USDOI). *U.S. Geological Survey (USGS). Landsat 8 (L8) Data Users Handbook (Version 1.0)*; EROS: Sioux Falls, SD, USA, 2015.

54. Nichol, E.J. High-Resolution Surface Temperature Patterns Related to Urban Morphology in a Tropical City: A Satellite-Based Study. *J. Appl. Meteorol.* **1996**, *35*, 135–146. [CrossRef]

55. Weng, Q.; Lu, D.; Schubring, J. Estimation of land surface temperature–vegetation abundance relationship for urban heat island studies. *Remote Sens. Environ.* **2004**, *89*, 467–483. [CrossRef]

56. Uni-Trend Inc. *The UNI-T®User Guidance Version 1.08.14s*; Uni-Trend Inc.: Shenzhen, China, 2014.

57. Jiménez-Muñoz, J.C.; Sobrino, J.A. A generalized single-channel method for retrieving land surface temperature from remote sensing data. *J. Geophys. Res. Atmos.* **2003**, *108*, 4688. [CrossRef]

58. NASA. Atmospheric Correction Parameter Calculator. Available online: http://atmcorr.gsfc.nasa.gov/ (accessed on 1 November 2019).

59. R Core Team. *R: A Language and Environment for Statistical Computing*; R Foundation for Statistical Computing: Vienna, Austria, 2019.

60. Bjørn-Helge Mevik and Ron Wehrens. The pls package: Principal component and partialleast squares regression in R. *J. Stat. Softw.* **2007**, *18*, 1–24.

61. Ghalanos, A.; Theussl, S. *Rsolnp: General Non-Linear Optimization Using Augmented Lagrange Multiplier Method. R Package Version 1.16*; R Foundation for Statistical Computing: Vienna, Austria, 2015.

62. Kim, J.H.; Gu, D.; Sohn, W.; Kil, S.H.; Kim, H.; Lee, D.K. Neighborhood Landscape Spatial Patterns and Land Surface Temperature: An Empirical Study on Single-Family Residential Areas in Austin, Texas. *Int. J. Environ. Re.s Public Health* **2016**, *13*, 880. [CrossRef] [PubMed]

63. Wong, M.S.; Nichol, J.E.; To, P.H.; Wang, J. A simple method for designation of urban ventilation corridors and its application to urban heat island analysis. *Build Environ.* **2010**, *45*, 1880–1889. [CrossRef]

64. Connors, J.P.; Galletti, C.S.; Chow, W.T.L. Landscape configuration and urban heat island effects: Assessing the relationship between landscape characteristics and land surface temperature in Phoenix, Arizona. *Landsc. Ecol.* **2012**, *28*, 271–283. [CrossRef]

65. Zhou, W.; Cao, F.; Wang, G. Effects of Spatial Pattern of Forest Vegetation on Urban Cooling in a Compact Megacity. *Forests* **2019**, *10*, 282. [CrossRef]

66. Zhang, S.-Y. Green Space System Planning in Shanghai. *Urban Plan Forum* **2002**, *6*, 14–16.

67. Cheng, J.; Yang, K.; Zhao, J.; Yuan, W.; Wu, J.-P. Variation of river system in center district of Shanghai and its impact factors during the last one hundred years. *Sci. Geogr. Sinica* **2007**, *27*, 85–91.

68. Wang, H.-J. Application of drainage mode in urban diked area: A new mode for urban drainage. *Shanghai Water* **2001**, *2*, 37–39.

69. Zhou, Z.-H. *Shanghai Historical Atlas*; Shanghai People's Publishing House: Shanghai, China, 1999.

70. Zhou, H.-M.; Gao, Y.; Ge, W.-Q.; Li, T.-T. The Research on the Relationship Between the Urban Expansion and the Change of the Urban Heat Island Distribution in Shanghai Area. *Ecol Environ.* **2008**, *17*, 163–168.

71. Zhang, H.; Qi, Z.-F.; Ye, X.-Y.; Cai, Y.-B.; Ma, W.-C.; Chen, M.-N. Analysis of land use/land cover change, population shift, and their effects on spatiotemporal patterns of urban heat islands in metropolitan shanghai, china. *Appl. Geogr.* **2013**, *44*, 121–133. [CrossRef]

72. Shanghai Forestry Bureau (SFB). Revisiting the Past 40-Year Environment in Shanghai. Available online: http://lhsr.sh.gov.cn/sites/ShanghaiGreen/dyn/ViewCon.ashx?ctgId=5b30d611-d8e6-467f-b36b-edb8a094bcb7&InfId=ffd2dbcd-5e7a-4222-aafb-c567f4ee7ac4 (accessed on 1 March 2019).

73. Chakraborty, S.D.; Kant, Y.; Mitra, D. Assessment of land surface temperature and heat fluxes over Delhi using remote sensing data. *J. Environ. Manag.* **2015**, *148*, 143–152. [CrossRef] [PubMed]

Article

Exploring the Synergies between Urban Overheating and Heatwaves (HWs) in Western Sydney

Hassan Saeed Khan [1,2,*], Riccardo Paolini [1], Mattheos Santamouris [1] and Peter Caccetta [2]

1   Faculty of Built Environment, University of New South Wales (UNSW), Sydney NSW 2052, Australia; r.paolini@unsw.edu.au (R.P.); m.santamouris@unsw.edu.au (M.S.)
2   Data-61, The Commonwealth Scientific and Industrial Research Organization (CSIRO), Underwood Ave, Floreat, Perth WA 6014, Australia; peter.caccetta@data61.csiro.au
*   Correspondence: hassan.khan@unsw.edu.au or hassan.khan@data61.csiro.au

Received: 3 December 2019; Accepted: 14 January 2020; Published: 18 January 2020

Abstract: There is no consensus regarding the change of magnitude of urban overheating during HW periods, and possible interactions between the two phenomena are still an open question, despite the increasing frequency and impacts of Heatwaves (HW). The purpose of this study is to explore the interactions between urban overheating and HWs in Sydney, which is under the influence of two synoptic circulation systems. For this purpose, a detailed analysis has been performed for the city of Sydney, while considering an urban (Observatory Hill), in the Central Business District (CBD), and a non-urban station in Western Sydney (Penrith Lakes). Summer 2017 was considered as a study period, and HW and Non-Heatwave (NHW) periods were identified to explore the interactions between urban overheating and HWs. A strong link was observed between urban overheating and HWs, and the difference between the peak average urban overheating magnitude during HWs and NHWs was around 8 °C. Additionally, the daytime urban overheating effect was more pronounced during the HWs when compared to nighttime. The advective flux was found as the most important interaction between urban overheating and HWs, in addition to the sensible and latent heat fluxes.

Keywords: heatwaves; urban overheating; urban heat island intensity; energy budget equation; sensible heat flux; latent heat flux; advective heat flux; Australian climatic conditions; coastal cities; desert climate

1. Introduction

Currently, urban areas around the world are hosting more than 50% of the world's population, and this figure is rapidly increasing [1]. Additionally, it is expected that, by 2025, 75% of the world's population will reside in coastal cities and, consequently, more built-up spaces will be required. Mostly, cities are hotter than the surrounding non-urban areas, because of the urban heat island (UHI) effect. Urban overheating is one of the most documented phenomena, affecting more than 400 cities around the world [2,3]. The major causes of urban overheating relate to synoptic weather conditions, thermal properties of the materials (absorbing solar radiations or opaque surfaces releasing heat), lack of vegetation, less evaporative surfaces, availability of heat sources or sinks in the cities, anthropogenic heat released in the cities, and the reduction of wind penetration due to the urban texture [4–8].

Urban overheating has a devastating impact on energy consumption, vulnerability, human health, economy, and environment, and these negative effects are further magnified during HWs. Urban overheating increases the building energy demand and the peak electricity demand [4]. In more detail, with every 1 °C rise in temperature, induced by urban overheating, peak electricity demand might increase by 0.45–4.6% [9]. Additionally, UHI might also induce up to 0.8 kWh/m²K of the average cooling energy penalty [10]. While forecasting the global energy demand, it was also concluded that UHI and other factors, such as population growth and rapid penetration of air-conditioning, may increase the cooling energy loads of residential and commercial buildings up to 275% and 750%, respectively, by 2050 [11].

In addition to the adverse effects on energy and economy, UHIs and HWs impact on human health is most frightening and is the matter of prime concern [12,13]. Worldwide, HWs are one of the leading causes of weather-related mortality [14]. A notable case is the European heatwave of 2003, which caused approximately 70,000 excess deaths [15]. In Australia, over the last two centuries, heat-related deaths exceeded those due to other climate hazards [16]. As in many studies, in Sydney, a significant increase in the mortality rate of elderly people, especially of women, was noted during HW episodes [17–20], with an increase in mortality rate by 13% computed for the summer of 2011 [20].

Heat-related mortality and morbidity also increase due to the combined effect of overheating in the cities and higher levels of pollutants, being also enhanced by high temperatures [21]. More in detail, a 4.5–12% increase in mortality rate was estimated in Sydney, due to a 10 °C rise in daily maximum temperatures and high concentrations of ozone and particulate matter [22], being notably contributed by photochemical smog formation [23,24]. People's adaptation to higher temperatures influences the mortality and morbidity rates and also the definition of the HW threshold temperatures towards the use of relative instead of absolute terms [25,26]. In particular, urban overheating impacts on low-income populations are also devastating, due to the poor thermal quality of dwellings, which worsens the exposure to extreme heat events and pollution. Resultantly, mortality, and morbidity ratios are higher in such urban areas [27].

Extreme heat events occur with higher frequency and intensity due to global climate change. For instance, HW frequency has already increased in the last few decades [28] and it is likely to increase further in the 21st century [29,30]. These regional and global changes are also affecting the UHI magnitude during the HWs. Cities are more vulnerable to extreme heat events, and HWs have more severe impacts due to their higher population density and additive UHI effect [31]. Recent findings on HWs, their causes, impact on energy, economy, and health, and its higher frequency have altered the research trend towards the identification of the potential interactions between UHIs and HWs, which affects the urban population [6,21,32].

Exploring the synergies between urban overheating and HWs helps in identifying the key factors that intensify the urban overheating, and more effectively plan for the application of mitigation technologies. No consensus exists on the relative change of the magnitude of urban overheating during HWs, and the interactions between both phenomena is still an open question [33–40]. Additionally, interactions between UHIs and HWs are mostly explored for noncoastal sites relative to the rural sites, and mostly positive interactions were noted at nighttime. For instance, in Beijing, China, the UHI is intensified at nighttime during HWs, and the difference between HW UHI and the background UHI is approximately 1.5 °C [35]. Similarly, higher nighttime UHI during the HWs was also found in Karachi, Shanghai, and Oklahoma [33,41,42]. No modification was noted in the UHI trend during the HWs in Philadelphia, PA, USA [43], while a reduction in UHI pattern during HWs was reported by Brazdil at Klementinum, Czech Republic [34]. In contrast to these studies, an amplified daytime UHI during HWs was found in Athens, while considering coastal stations as a reference [37]. Similarly, a positive daytime response between UHIs and HWs was noted in Shanghai, China, while considering the coastal reference station and the maximum difference between UHI during HWs and NHWs was around 1.5 °C [38].

In previous studies, either rural sites were compared with the urban sites or coastal sites were compared against the urban stations. However, the interaction between urban overheating and HWs in coastal cities close to a significant heat source such as desert has never been explored before. Sydney is located along the coastline of the South Pacific Ocean on the eastern side, while, on the west, it is exposed to advective airflow from one of the largest deserts in the world. Thus, it is under the influence of two different synoptic systems, (i) coastal winds and (ii) desert winds, as concluded in [5,44]. Synergistic interactions between UHIs and HWs also depend upon the boundary conditions of the investigated area. For instance, the urban-rural moisture contrast is considered as an essential interaction when noncoastal urban sites are compared with rural locations. In Baltimore, USA, during HWs, not only urban-rural temperature differences increased, but also the urban-rural moisture contrast, which is one of the major synergistic interactions between UHIs and HWs [41]. During HWs, typically, the sensible heat flux at the urban site increases [45], while, at the rural site, it is the latent heat flux that increases. This dynamic further intensifies the magnitude of the UHI during the HW [35]. High wind speed typically decreases the magnitude of UHIs [46]; however, advective heat flux from warm areas, increases the UHI and affects the interaction between UHIs and HWs, when coastal stations are considered [37].

The present study investigates the change in urban overheating magnitude during HWs in a coastal city, Sydney, being regionally affected by the desert and explores the synergistic interactions between urban overheating and HWs. Penrith Lakes in Western Sydney, influenced by the desert winds, at approximately 55 km from the coast, is taken as the nonurban (inland) station, while Observatory Hill located near the coast and Central Business District (CBD) Sydney is considered as urban reference station (coastal) to investigate the issue. The meteorological data of summer 2017 are analyzed to identify the HW and non-heatwave (NHW) periods. The specific energy budget contrast is computed in terms of sensible, latent, and advective heat fluxes to identify the potential interactions between urban overheating and HWs. To counteract urban overheating in Sydney, appropriate mitigation technologies can be devised by using the results of this study and, consequently, public health, energy savings, and environmental quality can be improved.

2. Methodology and Data

Sydney is the capital of the state of New South Wales (NSW), spreading along the south-eastern coastline, while, on the west, it is under the influence of one of the largest desert landforms in the world, the Australian arid biome [47]. From the eastern coastline to the western Blue Mountains, Sydney's metropolitan area extends by about 70 km [48]. For this study, Sydney is divided into three main zones, (i) Eastern Sydney, (ii) Inner Sydney, and (iii) Western Sydney. Eight different stations in three zones of Sydney are considered for the Heatwave (HW) calculation: Observatory Hill, Sydney Airport and Terrey Hills reserve in the eastern Sydney, Olympic park AWS and Canterbury Racecourse in the Inner Sydney, Horsley park Equestrian Centre AWS, Campbelltown and Penrith Lakes in Western Sydney (Figure 1). To quantify the magnitude of urban overheating and characterize the synoptic conditions in Sydney, Observatory Hill—which is highly influenced by the sea breeze—is considered as the reference urban (coastal) station in this study, as proposed in [5]. Penrith Lakes, which is influenced by the desert winds, is taken as a nonurban (inland) station, with among the highest maximum temperatures in Western Sydney, and Australia in general.

Weather data are obtained from the Australian Bureau of Meteorology (BOM) for the sites of Penrith Lakes, Observatory Hill (OBS Hill) and six other stations in eastern, western, and inner Sydney. Observatory Hill (151°12′18″ E, 33°51′39″ S, 39 m above mean sea level) is located on a hill covered by greenery, in the proximity to the coast and Sydney's central business district (CBD) [49] OBS Hill is in a mixed-use area, with industrial, commercial, and residential buildings, extending over 27 km^2, with a tree canopy cover of approximately 15.2% [50]. It is one of the densest areas in Sydney, with a population of 205,339, corresponding to 76.8 persons per hectare [51]. Wind is mostly blowing from sea to the site, and NE/SE winds are representing the sea breeze, while the winds from NW/SW blow from inner Sydney towards the reference station.

Penrith Lakes is in Western-Sydney, north-west of Observatory Hill, at about 55 km from the reference station and the nearest coastline is at 49 km distance. The weather station is located in Penrith Lakes regional park (150°40′42″ E, 33°43′10″ S, 24.7 m above mean sea level), with green areas, bare soil, and lakes in surroundings [49], while the close urban area of Penrith displays mostly residential and commercial buildings. Additionally, the Nepean river flows to the west of the station, and an aqua park and golf facility are also present nearby. The tree canopy cover of the area is around 25% of the total land area, much higher than Sydney CBD [50]. Temperatures at Penrith Lakes are quite high in summer and sometimes reaching 47 °C. The population of Penrith is approximately 13,000. Wind speed and direction vary and NE/SE winds represent the wind coming from inner Sydney to the site, while the NW/SW direction denotes the dry, warm, and downslope winds (Foehn wind)/desert wind, which might increase the site ambient temperatures.

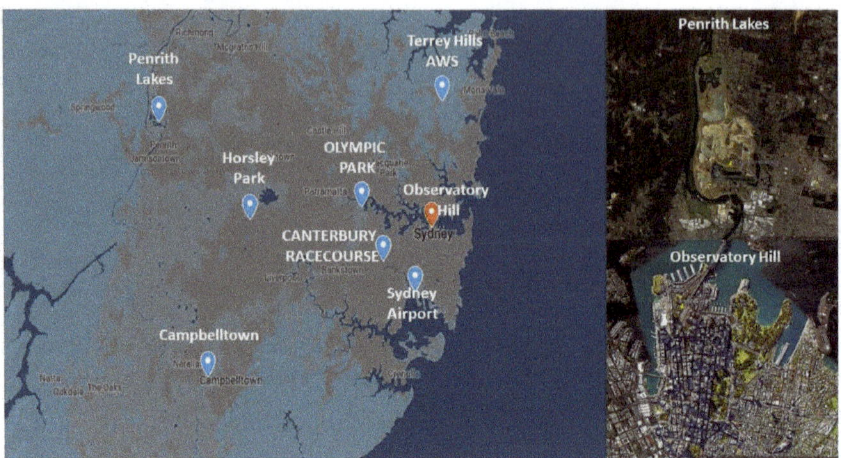

Figure 1. Selected weather stations and their locations.

The dataset that was composed of semi-hourly records was cleaned, and outliers and null-values were removed, while using the validation procedure that was proposed by Estevez et al. [52], comprising range tests, step tests, persistence tests, and relational tests. A two-step procedure was used to infill the gaps. The missing values up to three hours were linearly interpolated, while, for longer gaps, genetic algorithms have been applied to establish relations between the studied station and three nearest stations. Subsequently, the hourly averages were computed from 1999 to 2017. The dataset was analyzed to calculate the HW days and periods and understand the frequency and intensity of HW over the years. As done in [44], the period 1999–2017 was considered due to either the unavailability of data at some of the weather stations before the study period or issues with data quality and significant gaps, which could not be infilled for this type of analysis. The available meteorological parameters were the ambient temperature (near-surface air temperatures at 2 m height), Relative Humidity (RH), wind speed, and wind direction. Vapour pressure (Pv), saturated pressure (Ps), Absolute Humidity (AH), and dewpoint temperature were calculated from the available parameters. The summer of 2017 was selected as a study period, as it was one of the warmest summers in recent years, to identify the interactions between HWs & urban overheating at Penrith Lakes [44]. Additionally, a time series analysis of ambient temperature over Sydney from 1960 to 2016 demonstrated that the mean and maximum temperatures in summer have increased in recent years [53]. Monthly and seasonal UHI variations in Sydney have already been studied in [44], while the focus of the study is on urban overheating response during HWs.

2.1. HW Period Identification & UHII Calculation

Defining the HW period is not straightforward, as people in different areas have different climatic adaptability and, hence, the vulnerability factor might vary with the location. Consequently, there is no globally accepted definition of HW period, with multiple metrics being proposed in the literature [25,26,54,55]. The temperature threshold, spatial extension, and duration of the HW period are the most commonly used parameters for isolating HW events. Most of the metrics identify an HW when a threshold is trespassed by a given number of subsequent days with the limit being an absolute maximum temperature threshold (35 °C or 37 °C), or a relative maximum temperature (90th,

95th, or 97th percentile) [22,25,26]. The latter type of metrics consider percentiles instead of absolute temperatures to account for the climatic adaptability factor. There is also no consensus on the minimum number of subsequent days trespassing threshold to be considered an HW: two consecutive days are used in the USA, while two to five consecutive days in Europe and Australia are taken as the minimum duration [17,56–58]. Excess Heat Index (EHI) and Excess Heat Factors (EHF) also have been used to identify HW events, by considering the 95th percentile temperature threshold and three consecutive days [58]. In the present research, the temperature threshold is defined in terms of percentile, and the minimum duration is considered as three consecutive days, as in [37,55,56]. The number of HW days at Penrith Lakes, as calculated with 97th, 95th, and 90th percentiles were 208, 342, and 691 days, respectively, while the HW periods were 12, 31, and 84, respectively (Table 1).

Table 1. Number of Heatwave (HW) days and periods at Penrith Lakes and Observatory (OBS) Hill (1999–2017).

Weather Station	Threshold Temperature	Threshold Temperature (°C)	Number of HW Days	Number of HW Periods
OBS Hill (Reference)	97th Percentile	31.2	209	4
	95th Percentile	29.4	345	14
	90th Percentile	27.6	703	60
Penrith Lakes	97th Percentile	36.4	208	12
	95th Percentile	34.7	342	31
	90th Percentile	32	691	84

HWs is a regional phenomenon, thus, it was essential to know whether similar events were happening at other surrounding stations. Very few HW periods were noted at eight weather stations with the 97th percentile and three consecutive days definition. Additionally, these were not falling in the same period, which referred to the local overheating at a particular station. With the 95th percentile and three consecutive days definition, a common HW period at all weather stations was observed from 9–11th February, while the corresponding NHW period was considered from 13–15th February, as shown in Table 2. Two NHW periods of equal duration as of the HW period, one immediately before and one after the HW period, were considered in [37,41], to explore the synergies between urban overheating and HWs. However, in the present research, another HW period was noticed at Penrith Lakes from 4th–6th February (Table 2). Thus, only one HW and one NHW periods were considered for further analysis due to the gap of only two days between two HW periods. The threshold temperatures (95th percentile) were 34.7 °C and 29.4 °C, at Penrith Lakes and OBS Hill, respectively (Table 2). During the summer of 2017, a total of four HW periods were recorded at Penrith Lakes, two in January and two in February.

Table 2. Total Number of HW Periods and their duration in Summer 2017 (with 95th Percentile and three consecutive days definition).

Weather Station	Threshold Temperature (°C) (95th Percentile)	Number of HW Periods	Days for HW Periods in Summer 2017	Days for NHW Periods in Summer 2017
OBS Hill	29.4	2	29–31 Jan.	26–28 Jan.
			9–11 Feb.	13–15 Feb.
Penrith lakes	34.7	4	8–11 Jan.	4–7 Jan.
			16–18 Jan.	19–21 Jan.
			4–6 Feb.	1–3 Feb.
			9–11 Feb.	13–15 Feb.

The magnitude of urban overheating is computed as the ambient temperature difference between Penrith Lakes and OBS Hill ($\Delta T = T_{Penrith} - T_{OBS\ Hill}$). Thus, positive urban overheating magnitude indicates the higher temperatures at Penrith Lakes than in the City, while negative values mean higher temperatures at the coastal station. Mean urban overheating magnitude with standard deviations was also calculated for the whole HW and NHW periods to understand the variation in urban overheating patterns. Similarly, AH differences ($\Delta AH = AH_{Penrith\ Lakes} - AH_{OBS\ Hill}$) were also calculated between both sites, and a positive value denotes higher latent heat flux at Penrith Lakes, while a negative value shows higher latent heat flux at the coastal site.

2.2. Energy Budget Equation Exploration

The energy budget equation is used to understand the surface energy contrast between urban and rural sites. The quantification of the energy budget equation at both sites during HWs and NHWs helps in understanding the interactions between urban-overheating and HWs. In the energy budget equation ($Q^* + Q_F = Q_H + Q_E + \Delta Q_S + \Delta Q_A$, expressed in W/m^2), Q^* is the net all-wave radiative heat flux at earth surface. It consists of the balance of incoming and outgoing short-wave and long-wave radiations. Q_F is the anthropogenic heating flux, which is high at urban sites due to human activities, such as air-conditioning use. Q_H and Q_E are the turbulent sensible and latent heat fluxes, respectively, and their sum is also called available energy. The sensible and latent heat fluxes are mostly explored while investigating the interaction between UHIs and HWs and are also considered as one of the important reasons for urban-rural temperature difference [35]. Sensible heat flux is usually high at urban sites, mostly due to higher surface temperatures, while latent heat flux is high at the rural site, because of abundant site moisture. Urban-rural moisture contrast is considered as one of the important synergistic interactions between HWs & UHIs [41]. ΔQ_S is the storage heat flux, which is high at the urban site due to the higher heat storage capacity of surfaces. ΔQ_A is the advective heat flux. In this study, sensible and latent heat fluxes during HWs and NHWs were estimated through near-surface temperature (ambient temperatures) and absolute humidity values.

Typically, higher surface temperatures during the HWs increases the ambient temperatures, through convection and resultantly sensible heat flux [59]. Similarly, latent heat flux increases due to the higher evaporation rates at the site. Wind speed and direction helped in identifying the advective heat potential at both sites. In the present study, mainly sensible, latent, and advective heat fluxes are explored to identify the synergies between urban-overheating and HWs. Initially, analysis is made for one representative HW (11 February 2017) day, last day of the HW period, having maximum temperature, and one NHW day (14 February 2017), having minimum temperatures during considered NHW period. After a one-day analysis, it is repeated for all HW and NHW periods.

3. Results

The magnitude of urban overheating was computed during both HW and NHW periods. Urban-overheating magnitude at daytime was quite high (average peak around 10 °C at 17:00) during the HWs, as compared to the NHW period, especially in the afternoon (Figure 2).

The results show a strong connection between urban overheating and HWs. The specific pattern can be attributed either to advection from the sea breeze, cooling down the coastal station quickly during the day, and then intensified the urban overheating or because of the latent heat flux difference between both sites, got increased abruptly during the daytime.

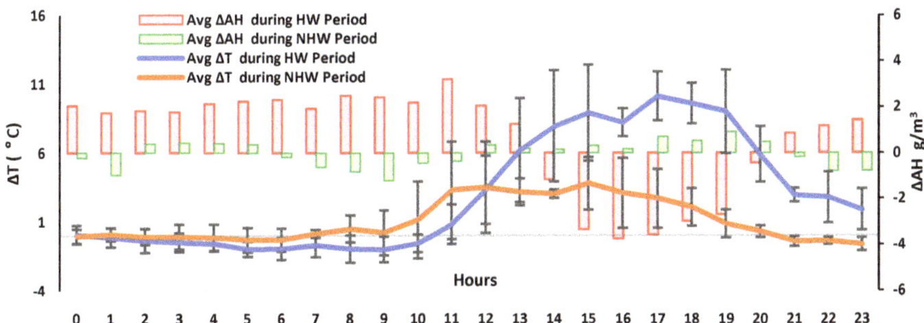

Figure 2. Mean urban overheating intensity (ΔT = Penrith Lakes − OBS Hill) and AH difference (ΔAH = Penrith Lakes − OBS Hill) during HW (9–11 February 2017) & NHW (13–15 February 2017) periods. The whiskers represent the standard deviation.

3.1. Coastal-Inland Moisture Contrast on Selected HW and NHW Days

During the HW day, the latent heat flux difference between the coastal and the inland sites increased in the daytime, compared to the NHW day (Figure 3). Generally, due to higher surface temperatures during the HW period, the evaporation rate increases, which increases the AH. The availability of vegetation and ground surface moisture enhances the evaporation/evapotranspiration rates. Typically, evapotranspiration/evaporation is more evident in rural areas, when compared to urban areas and, hence, these areas show lower ambient temperatures. When the AH difference, between both sites, during the HW period at daytime was increasing, the urban-overheating magnitude was also increasing (Figure 3). From here, it can be concluded that, despite having nonurban surfaces and being closer to the water bodies (lakes), latent heat flux at Penrith Lakes at daytime was lower, during the HW period when compared to the coastal site. The possible explanations include: (i) reduced evaporation rates at the site, (ii) advection of dry air at Penrith Lakes, (iii) advection of humid air from sea to the coastal site, and (iv) higher/constant evaporation or evapotranspiration at the coastal station. Temperature and AH profiles at Penrith Lakes and OBS Hill were explored further to understand the phenomena.

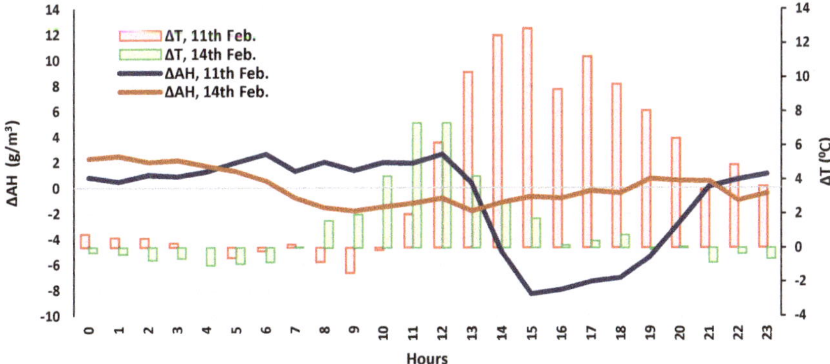

Figure 3. Absolute humidity difference (ΔAH) and urban-overheating (ΔT) between Penrith Lakes—OBS Hill during selected HW (11 February 2017) & NHW (14 February 2017) days.

3.2. Wind Speed and Direction on Selected HW and NHW Days

Wind speed and its directions were examined during HW and NHW days and observed the following. During HW day, at daytime wind speed at Penrith lakes increased from 2–5 m/s and it was blowing from desert to the site (Figure 4A). At this time, the wind was blowing from sea to site at OBS Hill, as shown in Figure 4C, and the urban-overheating magnitude was quite high. During the

NHW day, wind speed at Penrith lakes was slightly higher when compared to the HW day, and it was blowing from Inner-Sydney to the site (Figure 4B), and urban overheating magnitude was quite low, as compared to the HW day. The nocturnal wind direction at Penrith lakes during HW day (desert to the site) and NHW days (Inner Sydney to the site) did not change much when compared to the daytime. However, windspeed also got reduced at nighttime and urban overheating magnitude.

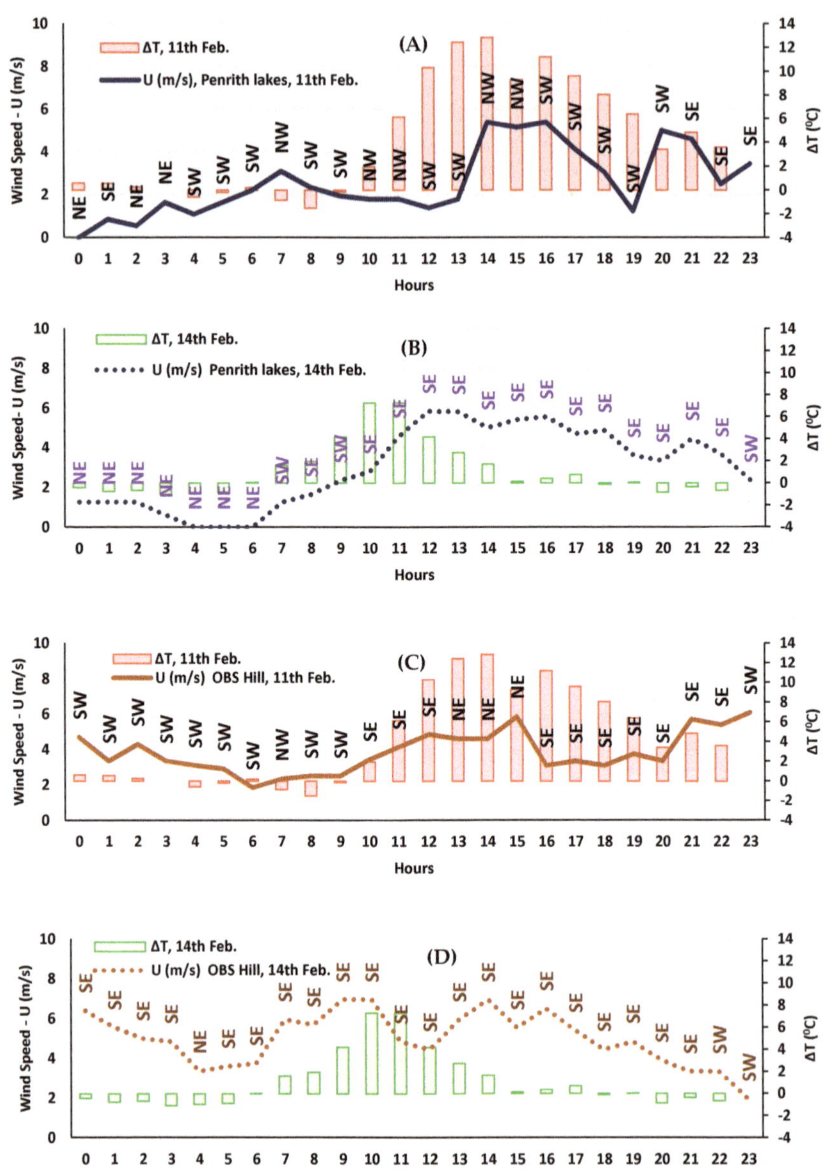

Figure 4. Wind speed-U (m/s) and direction comparison with urban overheating (ΔT) at Penrith Lakes and OBS Hill during selected HW (11 February 2017) and NHW (14 February 2017) days. (**A**): Penrith Lakes during HW day, (**B**) Penrith Lakes during NHW day, (**C**) OBS Hill during HW day, (**D**) OBS Hill during NHW day.

On the other hand, at OBS Hill during the HW day, the nocturnal wind direction was from inland to the site, while during the NHW day wind was blowing from sea to the site (Figure 4D). Wind speed during the NHW day at OBS Hill was much higher when compared to the HW day.

3.3. Absolute Humidity and UHI Magnitude at Night-Time

During HW & NHW days, AH at Penrith lakes was higher at nighttime, when compared to OBS Hill (Figure 5B). The land cover (retaining more moisture content), proximity to the lakes, wind speed, and direction and evaporation of surface moisture (formed due to condensation) can explain the higher AH and lower nocturnal temperature at Penrith Lakes. The principal reason could relate to non-urban surfaces, having higher evaporation potential and increasing the latent heat flux at the site. Precipitation could be another important reason for higher soil moisture at Penrith Lakes, which was observed on the 7th and 8th of February.

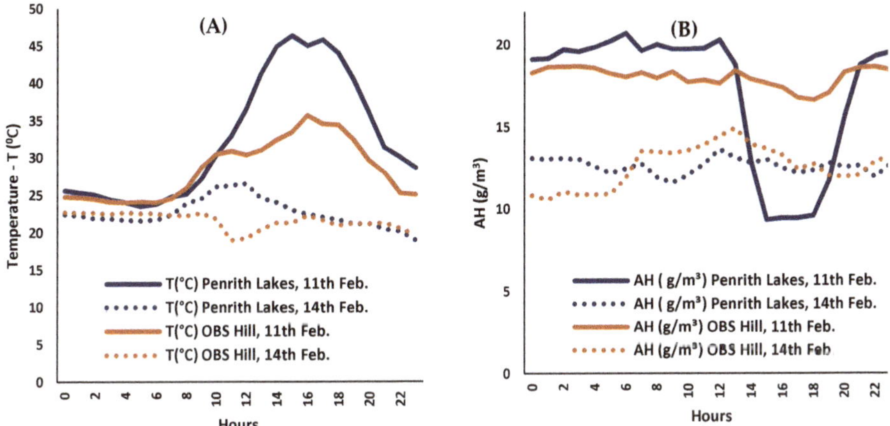

Figure 5. Temperature T (°C) and AH (g/m^3) profiles at Penrith Lakes & OBS Hill on selected HW (11 February 2017) & NHW (14 February 2017) days. (**A**) Temperature profiles in °C at Penrith Lakes and OBS Hill on HW and NHW days, (**B**) AH profiles in g/m^3 at Penrith Lakes and OBS Hill on HW and NHW days.

Dew formation is another potential reason for the increased AH value in the morning, as, in early morning, warm soil can evaporate the excessive surface moisture, which formed due to condensation. This phenomenon occurs if the surface temperatures are less than or equal to the dew point temperature, then humidity from the air condenses, and dew is formed. Site ambient temperatures at Penrith Lakes and OBS Hill were compared with dewpoint temperatures during both HW and NHW periods to explore the chances of dew formation, and it was found that, at Penrith Lakes, there were chances of dew formation during HW days, as shown in Figure 6A. Thus, a sharp increase in AH value at Penrith lakes at 6 am could be because of the evaporation of surface moisture.

Nighttime temperatures at both stations are almost the same during both HW and NHW days (Figure 5A). At OBS Hill, steady sea surface temperatures penetrating at the site are maintaining the air temperatures and AH within a specific range, while the radiative cooling process at Penrith lakes due to its non-urban surfaces is quickly reducing ambient temperatures. Thus, the nocturnal urban-overheating magnitude during both HWs and NHWs is negligible.

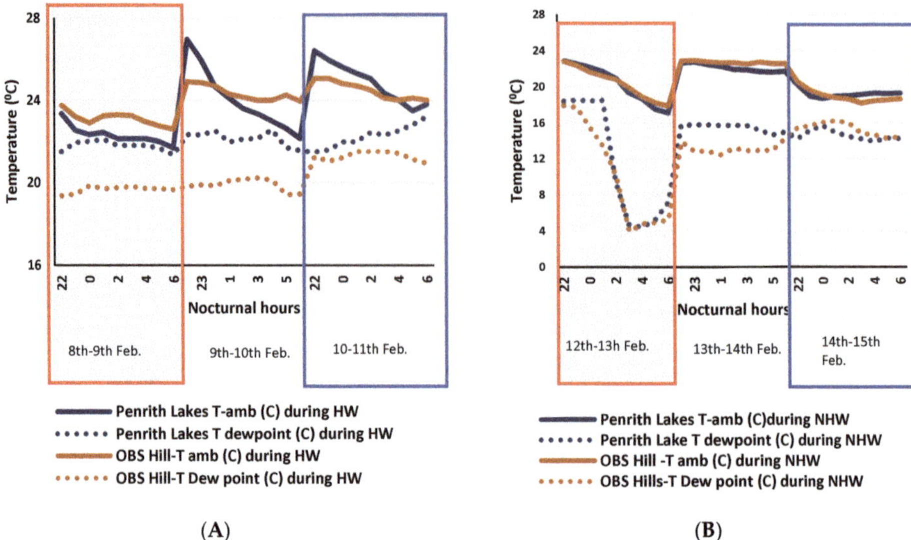

(**A**)   (**B**)

Figure 6. Nocturnal ambient temperatures & dew point temperatures at Penrith Lakes & OBS Hill during HW (9–11 February 2017) and NHW periods. (**A**) during the HW and (**B**) NHW period.

3.4. Daytime Urban Overheating Magnitude

Daytime temperatures at Penrith lakes were much higher when compared to OBS Hill during HW day (Figure 5A). During this time, AH at the site got reduced very quickly and then started increasing again in the evening (Figure 5B). Until midday, the AH was quite high and consistent at the site, but a sharp reduction in AH value referred to the change in wind speed and direction. During the whole HW day, wind at Penrith lakes was blowing from desert to the site; however, until 1 p.m., wind speed was quite low (Figure 4A). Advection from the desert wind with higher wind speed after 1 p.m., making the evaporation/evapotranspiration process ineffective at the site and, resultantly, drastic increase in temperature was noted at Penrith Lakes. Wind speed started decreasing at 4 p.m., and resultantly temperature started falling and AH increased again because of the restoration of the evaporation process.

The analysis of the data at Penrith Lakes during the whole HW period (9th–11th Feb.) & NHW period (13th–15th Feb.) found that daytime temperatures were consistently increasing at Penrith lakes from first HW day to the last HW day (Figure 7A). On the other hand, latent heat flux during daytime at the site was getting reduced similarly, as shown in Figure 7C. The wind blowing from inner Sydney to the site (see methodology and data) is helping to reduce the site temperatures as site evaporation potential is not getting compromised (Figure 8A,B). On the other hand, when the wind is blowing from desert to the site (Figure 8A), ambient temperatures increase (Figure 7A), as the evaporation process is becoming ineffective (Figure 7C). It is due to dry, warm air, which is causing the fast depletion of available moisture and decreasing the latent heat flux at Penrith Lakes. Additionally, as wind speed from the desert is increasing, AH is getting further reduced and drastically increasing the temperatures.

In contrast to Penrith Lakes, daytime AH at OBS Hill, during both HW & NHW days, was high (Figure 7D) and consistent and temperatures were also low in the same proportion (Figure 7B). This is due to the advection of humid air from the sea as during both HW & NHW days, the wind was blowing from sea to OBS Hill (Figure 4C,D). However, the wind speed was much higher, as compared to the HW day, during the NHW day. Thus, the temperatures at OBS Hill were accordingly low (Figure 5A).

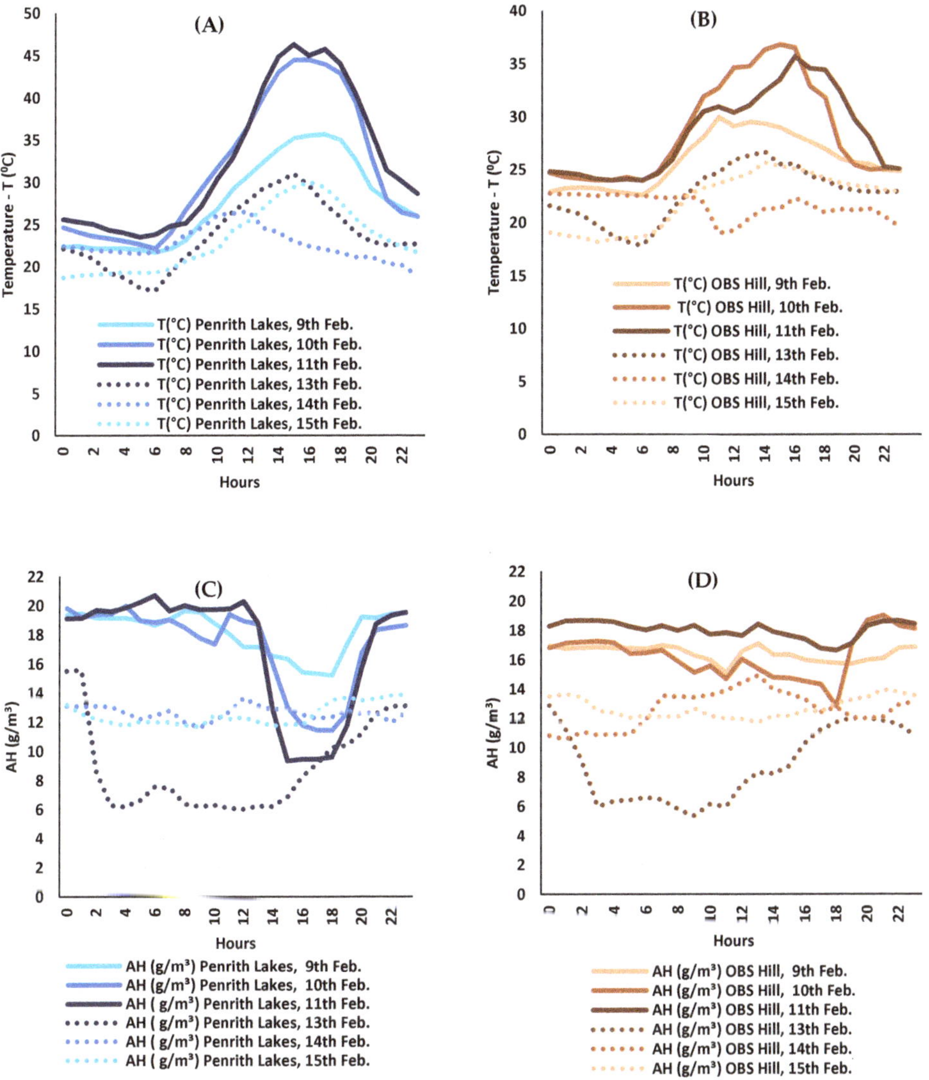

Figure 7. Temperature profile (°C) and AH (g/m^3) at Penrith Lakes and OBS Hill during HW (9–11 February 2017) & NHW (13–15 February 2017) Periods. (**A**) Temperature profiles at Penrith Lakes during HW and NHW periods, (**B**) Temperature profiles at OBS Hill during HW and NHW periods, (**C**) AH profiles at Penrith Lakes during HW and NHW periods, (**D**) AH profiles at OBS Hill during HW and NHW periods.

On the other hand, during the HW day, although the temperatures were relatively high at OBS Hill, but because of higher evaporation from the sea, AH was also higher. After analyzing the three consecutive HW (9–11th Feb.) and NHW (13–15th Feb.) days at OBS Hill, it was found that, during both HW and NHW periods in the daytime, the wind was blowing from sea to the site (Figure 8C,D). However, as wind speed was getting reduced, AH was also getting reduced due to less advection from the sea and resultantly temperatures were increasing (Figure 7B,D). When the wind was blowing from inland to the coastal site, AH was reduced, and temperatures increased.

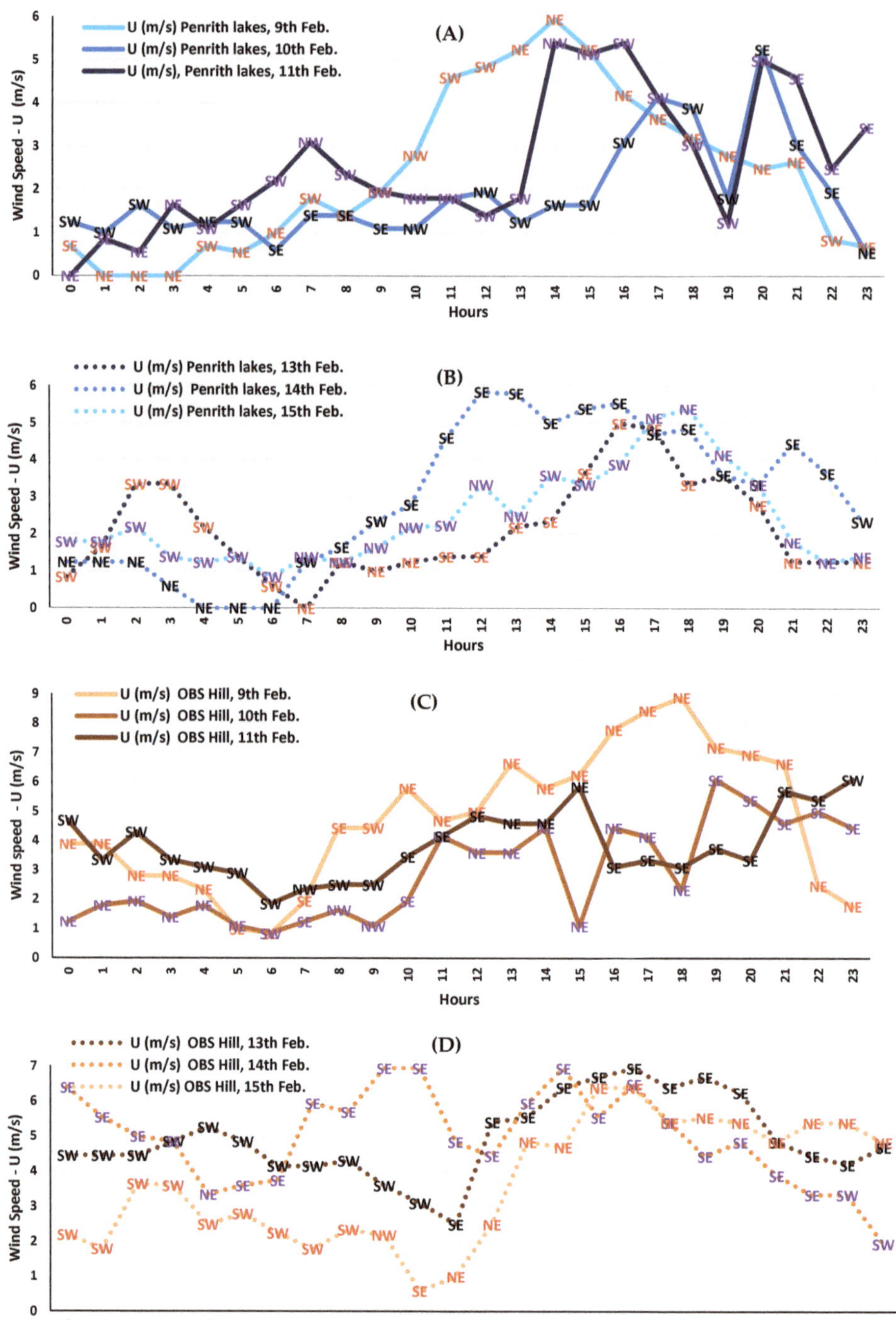

Figure 8. Wind Speed (m/sec) & wind direction at Penrith Lakes and OBS Hill during HW (9–11 February 2017) and NHW (13–15 February 2017) periods. (**A**) Penrith Lakes during the HW period, (**B**) Penrith Lakes during NHW period, (**C**) OBS Hill during HW period, (**D**) OBS Hill during the NHW period.

4. Discussion

The urban population is at higher thermal risk due to rapid urbanization and extreme heat events. Hence, comprehensive knowledge is required for identifying the factors, magnifying the urban-overheating, and changing the urban-rural energy contrast. In the present study, the very first time, the impact of extreme heat events on urban overheating is measured in a coastal city, which is also under the influence of the desert climate. The results reveal that the urban overheating effect gets amplified during the HWs, in contrast to some of the previous studies, where no modification or reduction in the UHI pattern was noticed during the HWs [34,43].

Additionally, it is found that the urban overheating effect is more pronounced in the daytime, especially in the afternoon. These results are consistent with the studies that were carried out in the coastal cities [37,38]. However, these results differ from other studies, where rural and noncoastal sites were compared, and due to a quick radiative cooling process at rural stations and slow cooling at urban sites, the nocturnal UHI effect was more intense during the HWs [33,35,41,42,60]. The nighttime urban overheating effect, in the present study, during HWs is unnoticeable. It is due to the rapid radiative cooling process at Penrith Lakes, due to non-urban surfaces and maintained temperatures at OBS Hill due to a steady sea breeze.

The difference between the peak average magnitude of urban overheating during HWs and NHWs is approximately 8 °C. In previous studies, with coastal or non-coastal reference stations, the maximum difference between HW and NHW urban overheating was from 1–2.5 °C [33,35,37,38]. The greater magnitude of urban overheating in Sydney is due to the presence of two different synoptic systems (heating mechanism—desert winds and cooling mechanism—coastal winds), on both sides of the city, which makes western Sydney more vulnerable. In the present study, advective heat flux is found as the dominant interaction between urban overheating and HWs, in addition to the sensible and latent heat fluxes. Advective heat flux was considered to be inconsequential for noncoastal cities; however, in coastal cities, it was one of the primary reasons for the intensification of urban overheating during HWs [37,38]. In noncoastal cities, urban-rural moisture contrast was considered a major interaction between UHIs and HWs [35,37,41].

Moreover, synergies between UHIs and HWs may get diminished if there is an insufficient latent heat flux difference between urban and rural sites [35]. In France, extreme temperatures were noted in 2003, because of the reduced evapotranspiration and latent cooling, which were linked to the soil moisture conditions [61]. Higher latent heat flux differences between urban and rural areas were also considered as the most important synergistic interaction in Beijing, China [35]. However, in the present study, sensible and latent heat fluxes at Penrith Lakes are also getting affected by the winds. Desert winds from the west are causing a fast depletion of available moisture at Penrith Lakes, consequently decreasing the total evaporation/evapotranspiration and, thus, the latent heat flux, while increasing the ambient temperatures. On the other side, a steady sea breeze at the coastal site during the HW event maintained the temperatures and humidity at a certain level. Generally, higher wind speed reduces the magnitude of UHIs [2,62,63]; however, in this case, winds with higher speed from the desert are increasing the temperatures at the inland station and, consequently, urban overheating. The impact of wind speed and direction (coastal and desert wind) on urban overheating at the coastal station highlights the importance of advective heat flux and shows it as the most influential interaction between urban overheating and HWs in Sydney.

Western Sydney's population is at high thermal risk, due to higher outdoor temperatures during extreme heatwave events. Thus, the chances of heat-related mortality, morbidity, air pollution, and electricity consumption increase [6]. Not only appropriate mitigation technologies need to be employed in Western Sydney, but CBD and eastern Sydney should be mitigated to allow for the sea breeze penetration in the Western Sydney through ventilation corridors to maintain the outdoor thermal comfort. Thus, through proper urban planning initiatives and design, excess urban heat can be mitigated and, consequently, it can help in reducing the economic burden from the health and energy sector. A detailed investigation of these mitigation technologies will be carried out in a subsequent paper. These results are not only applicable in Sydney, but they can also be applied to other cities, having similar boundary conditions, such as Muscat (Oman), Dubai(U.A.E.), Jeddah (Saudi Arabia), Dammam (Saudi Arabia), Casablanca (Morocco), etc. [2,64,65].

Reliance on the data from a single summer/year to identify the interaction between urban overheating and HWs is one of the limitations of this study, and results can be validated by considering data from some more recent summers. Additionally, this study can be further enhanced by analyzing the other weather stations in inner and Western Sydney. Other components of the energy budget equation may also be explored to understand the interactions between urban overheating and HWs further.

5. Conclusions

The frequency and intensity of HWs are increasing in the 21st century, negatively affecting the outdoor thermal comfort in urban areas. This study focuses on the interactions between urban overheating and HWs in a coastal city (Sydney, Australia), which is affected by a desert landmass. The results show a positive response between urban overheating and HWs, which gets magnified during the daytime, instead of night-time, consistently with the literature. Additionally, the difference between peak average urban overheating magnitude during HWs and NHWs was around 8 °C, which exceeds the most commonly documented figures. The greater difference between the magnitude of urban overheating during HWs and NHWs is attributed to the presence of two synoptic systems, which become further active during HWs. The increased severity of urban climate during HWs might severely affect human health, environmental quality by increasing the pollutants in the atmosphere, and economy by increasing the energy demand. Urban-rural moisture contrast is considered as one of the leading factors, intensifying the UHI magnitude during the HWs. However, advective heat flux was found as the dominant interaction between both these phenomena in addition to sensible and latent heat fluxes. Further, the advective heat flux also affected the sensible and latent heat fluxes at both sites and intensified the urban overheating. The results will help to devise the appropriate strategies to mitigate extreme heat events and design countermeasures to urban overheating.

Author Contributions: All authors (H.S.K., R.P., M.S., and P.C.) contributed equally in conceptualization, methodology, analyzing the data sets and preparation of the manuscript. All authors have read and agreed to the published version of the manuscript

Funding: The authors kindly acknowledge the Australian Bureau of Meteorology for the weather data, analyzed in this study, whose post-processing and validation was initially supported by Sydney Water and CRC for Low Carbon Living with the research contract 'SP0012: Strategic Study on the Cooling Potential and Impact of Urban Climate Mitigation Techniques in Western Sydney'. The data analysis was also supported by the City of Parramatta, with the research contract 'Parramatta Urban Overheating'.

Acknowledgments: Hassan Saeed Khan thankfully acknowledges UNSW Sydney and Data 61 CSIRO for supporting his Ph.D. in the form of University international postgraduate award and top-up scholarships.

Conflicts of Interest: The authors do not have any conflict of interest.

References

1. United Nation. *UN Finds World' s Population Is Increasingly Urban with More Than Half Living in Urban Areas Today and Another 2.5 Billion Expected by 2050 [Press Release]*; United Nation: New York, NY, USA, 2014.

2. Santamouris, M. Analyzing the heat island magnitude and characteristics in one hundred Asian and Australian cities and regions. *Sci. Total Environ.* **2015**, *512–513*, 582–598. [CrossRef] [PubMed]

3. Santamouris, M. Innovating to zero the building sector in Europe: Minimising the energy consumption, eradication of the energy poverty and mitigating the local climate change. *Sol. Energy* **2016**, *128*, 61–94. [CrossRef]

4. Santamouris, M. facing the problem of overheating in Australian cities. *Archit. Aust.* **2019**, *108*, 54–56.

5. Santamouris, M.; Haddad, S.; Fiorito, F.; Osmond, P.; Ding, L.; Prasad, D.; Zhai, X.; Wang, R. Urban heat island and overheating characteristics in Sydney, Australia. An analysis of multiyear measurements. *Sustainability* **2017**, *9*, 712. [CrossRef]

6. Santamouris, M. Recent Progress on Urban Overheating and Heat Island Research. Integrated Assessment of the Energy, Environmental, Vulnerability and Health Impact Synergies with the Global Climate Change. *Energy Build.* **2019**, *15*, 109482.

7. Khan, H.S.; Asif, M. Impact of green roof and orientation on the energy performance of buildings: A case study from Saudi Arabia. *Sustainability* **2017**, *9*, 640. [CrossRef]

8. Khan, H.; Asif, M.; Mohammed, M. Case Study of a Nearly Zero Energy Building in Italian Climatic Conditions. *Infrastructures* **2017**, *2*, 19. [CrossRef]

9. Santamouris, M.; Cartalis, C.; Synnefa, A.; Kolokotsa, D. On the impact of urban heat island and global warming on the power demand and electricity consumption of buildings—A review. *Energy Build.* **2015**, *98*, 119–124. [CrossRef]

10. Santamouris, M. On the energy impact of urban heat island and global warming on buildings. *Energy Build.* **2014**, *82*, 100–113. [CrossRef]

11. Santamouris, M. Cooling the buildings–past, present and future. *Energy Build.* **2016**, *128*, 617–638. [CrossRef]

12. Patz, J.A.; Campbell-Lendrum, D.; Holloway, T.; Foley, J.A. Impact of regional climate change on human health. *Nature* **2005**, *438*, 310. [CrossRef] [PubMed]

13. Basu, R.; Samet, J.M. Relation between elevated ambient temperature and mortality: A review of the epidemiologic evidence. *Epidemiol. Rev.* **2002**, *24*, 190–202. [CrossRef] [PubMed]

14. De Ridder, K.; Maiheu, B.; Lauwaet, D.; Daglis, I.; Keramitsoglou, I.; Kourtidis, K.; Manunta, P.; Paganini, M. Urban Heat Island Intensification during Hot Spells—The Case of Paris during the Summer of 2003. *Urban Sci.* **2016**, *1*, 3. [CrossRef]

15. Robine, J.M.; Cheung, S.L.K.; Le Roy, S.; Van Oyen, H.; Griffiths, C.; Michel, J.P.; Herrmann, F.R. Death toll exceeded 70,000 in Europe during the summer of 2003. *Comptes Rendus Biol.* **2008**, *331*, 171–178. [CrossRef] [PubMed]

16. National Climate Change Adaptation Research Facility. *Managing Heatwave Impacts under Climate Change—Policy Guidance Brief 9*; National Climate Change Adaptation Research Facility: Gold Coast, Australia, 2013; Volume 6.

17. Tong, S.; Wang, X.Y.; Yu, W.; Chen, D.; Wang, X. The impact of heatwaves on mortality in Australia: A multicity study. *BMJ Open* **2014**, *4*, 1–6. [CrossRef]

18. Vaneckova, P.; Hart, M.A.; Beggs, P.J.; De Dear, R.J. Synoptic analysis of heat-related mortality in Sydney, Australia, 1993–2001. *Int. J. Biometeorol.* **2008**, *52*, 439–451. [CrossRef]

19. Vaneckova, P.; Beggs, P.J.; Jacobson, C.R. Spatial analysis of heat-related mortality among the elderly between 1993 and 2004 in Sydney, Australia. *Soc. Sci. Med.* **2010**, *70*, 293–304. [CrossRef]

20. Schaffer, A.; Muscatello, D.; Broome, R.; Corbett, S.; Smith, W. Emergency department visits, ambulance calls, and mortality associated with an exceptional of in Sydney, Australia, 2011: A time-series analysis. *Environ. Heal. A Glob. Access Sci. Source* **2012**, *11*, 1–8. [CrossRef]

21. Paravantis, J.; Santamouris, M.; Cartalis, C.; Efthymiou, C. Mortality Associated with High Ambient Temperatures, Heatwaves, and the Urban Heat Island in Athens, Greece. *Sustainability* **2017**, *9*, 606. [CrossRef]

22. Vaneckova, P.; Beggs, P.J.; de Dear, R.J.; McCracken, K.W.J. Effect of temperature on mortality during the six warmer months in Sydney, Australia, between 1993 and 2004. *Environ. Res.* **2008**, *108*, 361–369. [CrossRef]

23. Lai, L.W.; Cheng, W.L. Air quality influenced by urban heat island coupled with synoptic weather patterns. *Sci. Total Environ.* **2009**, *407*, 2724–2733. [CrossRef] [PubMed]

24. Lebassi, B.; Gonzalez, J.; Fabris, D.; Maurer, E.; Miller, N.; Milesi, C.; Switzer, P.; Bornstein, R. Observed 1970–2005 cooling of summer daytime temperatures in coastal California. *J. Clim.* **2009**, *22*, 3558–3573. [CrossRef]

25. Tong, S.; FitzGerald, G.; Wang, X.Y.; Aitken, P.; Tippett, V.; Chen, D.; Wang, X.; Guo, Y. Exploration of the health risk-based definition for heatwave: A multi-city study. *Environ. Res.* **2015**, *142*, 696–702. [CrossRef] [PubMed]

26. Brooke Anderson, G.; Bell, M.L. Heat waves in the United States: Mortality risk during heat waves and effect modification by heat wave characteristics in 43 U.S. communities. *Environ. Health Perspect.* **2011**, *119*. [CrossRef]

27. Kolokotsa, D.; Santamouris, M. Review of the indoor environmental quality and energy consumption studies for low income households in Europe. *Sci. Total Environ.* **2015**, *536*, 316–330. [CrossRef]

28. Peterson, T.C.; Heim, R.R.; Hirsch, R.; Kaiser, D.P.; Brooks, H.; Diffenbaugh, N.S.; Dole, R.M.; Giovannettone, J.P.; Guirguis, K.; Karl, T.R.; et al. Monitoring and understanding changes in heat waves, cold waves, floods, and droughts in the United States: State of knowledge. *Bull. Am. Meteorol. Soc.* **2013**, *94*, 821–834. [CrossRef]

29. Coumou, D.; Robinson, A. Historic and future increase in the global land area affected by monthly heat extremes. *Environ. Res. Lett.* **2013**, *8*, 034018. [CrossRef]

30. Stocker, T. *IPCC Summary for Policymakers in Climate Change 2013: The Physical Science Basis*; Cambridge University Press: Cambridge, UK; New York, NY, USA, 2013.

31. Schatz, J.; Kucharik, C.J. Urban climate effects on extreme temperatures in Madison, Wisconsin, USA. *Environ. Res. Lett.* **2015**, *10*, 094024. [CrossRef]

32. Mitchell, D.; Heaviside, C.; Vardoulakis, S.; Huntingford, C.; Masato, G.; Guillod, B.P.; Frumhoff, P.; Bowery, A.; Wallom, D.; Allen, M. Attributing human mortality during extreme heat waves to anthropogenic climate change. *Environ. Res. Lett.* **2016**, *11*, 074006. [CrossRef]

33. Rizvi, S.H.; Alam, K.; Iqbal, M.J. Spatio-temporal variations in urban heat island and its interaction with heat wave. *J. Atmos. Solar-Terrestrial Phys.* **2019**, *185*, 50–57. [CrossRef]

34. Brázdil, R.; Budíková, M. An urban bias in air temperature fluctuations at the Klementinum, Prague, the Czech Republic. *Atmos. Environ.* **1999**, *33*, 4211–4217. [CrossRef]

35. Li, D.; Sun, T.; Liu, M.; Yang, L.; Wang, L.; Gao, Z. Contrasting responses of urban and rural surface energy budgets to heat waves explain synergies between urban heat islands and heat waves. *Environ. Res. Lett.* **2015**, *10*, 054009. [CrossRef]

36. Cheval, S.; Dumitrescu, A.; Bell, A. The urban heat island of Bucharest during the extreme high temperatures of July 2007. *Theor. Appl. Climatol.* **2009**, *97*, 391–401. [CrossRef]

37. Founda, D.; Santamouris, M. Synergies between Urban Heat Island and Heat Waves in Athens (Greece), during an extremely hot summer (2012). *Sci. Rep.* **2017**, *7*, 1–11. [CrossRef]

38. Ao, X.; Wang, L.; Zhi, X.; Gu, W.; Yang, H.; Li, D. Observed Synergies between Urban Heat Islands and Heat Waves and Their Controlling Factors in Shanghai, China. *J. Appl. Meteorol. Climatol.* **2019**, *58*, 1955–1972. [CrossRef]

39. Zhao, L.; Lee, X.; Smith, R.B.; Oleson, K. Strong contributions of local background climate to urban heat islands. *Nature* **2014**, *511*, 216–219. [CrossRef]

40. Zhao, L.; Oppenheimer, M.; Zhu, Q.; Baldwin, J.W.; Ebi, K.L.; Bou-Zeid, E.; Guan, K.; Liu, X. Interactions between urban heat islands and heat waves. *Environ. Res. Lett.* **2018**, *13*, 034003. [CrossRef]

41. Li, D.; Bou-Zeid, E. Synergistic interactions between urban heat islands and heat waves: The impact in cities is larger than the sum of its parts. *J. Appl. Meteorol. Climatol.* **2013**, *52*, 2051–2064. [CrossRef]

42. Basara, J.B.; Basara, H.G.; Illston, B.G.; Crawford, K.C. The Impact of the Urban Heat Island during an Intense Heat Wave in Oklahoma City. *Adv. Meteorol.* **2010**, *2010*. [CrossRef]

43. Ramamurthy, P.; Bou-Zeid, E. Heatwaves and urban heat islands: A comparative analysis of multiple cities. *J. Geophys. Res.* **2017**, *122*, 168–178. [CrossRef]

44. Yun, G.Y.; Ngarambe, J.; Duhirwe, P.N.; Ulpiani, G.; Paolini, R.; Haddad, S.; Vasilakopoulou, K.; Santamouris, M. Predicting the magnitude and the characteristics of the urban heat island in coastal cities in the proximity of desert landforms. The case of Sydney. *Sci. Total Environ.* **2020**, *709*, 136068. [CrossRef] [PubMed]

45. Grimmond, S. Urbanization and global environmental change: Local effects of urban warming. *Geogr. J.* **2007**, *173*, 83–88. [CrossRef]

46. Arnfield, A.J. Two decades of urban climate research: A review of turbulence, exchanges of energy and water, and the urban heat island. *Int. J. Climatol.* **2003**, *23*, 1–26. [CrossRef]

47. Byrne, M.; Yeates, D.K.; Joseph, L.; Kearney, M.; Bowler, J.; Williams, M.A.J.; Cooper, S.; Donnellan, S.C.; Keogh, J.S.; Leys, R.; et al. Birth of a biome: Insights into the assembly and maintenance of the Australian arid zone biota. *Mol. Ecol.* **2008**, *17*, 4398–4417. [CrossRef]

48. Australian Bureau of Statistics. Australian Statistical Geography Standard: Volume 4—Significant Urban Areas. *Urban Centres Localities Section State* **2011**, *4*, 29–31.

49. Australian Government Bureau of Meteorology Climate Data Online. Available online: http://www.bom.gov.au/climate/data/index.shtml?bookmark=201 (accessed on 15 August 2019).

50. Jacbos, B.; Mikhailovich, N.; Delaney, C. *Benchmarking Australia's Urban Tree Canopy: An i-Tree Assessment, Final Report 2014*; Institute for Sustainable Futures: Sydney, Australia, 2014; Volume 47.

51. Australian Bureau of Statistics. *Regional Population Growth, Australia, 2016-17-Population Density*; Australian Bureau of Statistics: Canberra, Australia, 2017.

52. Estévez, J.; Gavilán, P.; Giráldez, J.V. Guidelines on validation procedures for meteorological data from automatic weather stations. *J. Hydrol.* **2011**, *402*, 144–154. [CrossRef]

53. Livada, I.; Synnefa, A.; Haddad, S.; Paolini, R.; Garshasbi, S.; Ulpiani, G.; Fiorito, F.; Vassilakopoulou, K.; Osmond, P.; Santamouris, M. Time series analysis of ambient air-temperature during the period 1970–2016 over Sydney, Australia. *Sci. Total Environ.* **2019**, *648*, 1627–1638. [CrossRef]

54. Barnett, A.G.; Hajat, S.; Gasparrini, A.; Rocklöv, J. Cold and heat waves in the United States. *Environ. Res.* **2012**, *112*, 218–224. [CrossRef]

55. Pyrgou, A. Synergy of Urban Heat Island and Heat Waves. Ph.D. Thesis, The Cyprus Institute, Nicosia, Cyprus, 2018.

56. Perkins, S.E.; Alexander, L.V. On the measurement of heat waves. *J. Clim.* **2013**, *26*, 4500–4517. [CrossRef]

57. Kuglitsch, F.G.; Toreti, A.; Xoplaki, E.; Della-Marta, P.M.; Zerefos, C.S.; Trke, M.; Luterbacher, J. Heat wave changes in the eastern mediterranean since 1960. *Geophys. Res. Lett.* **2010**, *37*. [CrossRef]

58. Nairn, J.; Fawcett, R. *Defining Heatwaves: Heatwave Defined as a Heat-Impact Event Servicing all Communiy and Business Sectors in Australia*; CAWCR Technical Report No 060; The Centre for Australian Weather and Climate Research A partnership between the Bureau of Meteorology and CSIRO: Kent Town, South Australia, 2013.

59. Guan, K. Surface and ambient air temperatures associated with different ground material: A case study at the University of California, Berkeley. *Surf. Air Temp. Gr. Mater.* **2011**, *196*, 1–14.

60. Li, D.; Sun, T.; Liu, M.; Wang, L.; Gao, Z. Changes in wind speed under enhance urban heat islands in the Beijing metropolitan area. *J. Appl. Meteorol. Climatol.* **2016**, *55*, 2369–2375. [CrossRef]

61. Fischer, E.M.; Seneviratne, S.I.; Vidale, P.L.; Lüthi, D.; Schär, C. Soil moisture-atmosphere interactions during the 2003 European summer heat wave. *J. Clim.* **2007**, *20*, 5081–5099. [CrossRef]

62. Camilloni, I.; Barrucand, M. Temporal variability of the Buenos Aires, Argentina, urban heat island. *Theor. Appl. Climatol.* **2012**, *107*, 47–58. [CrossRef]

63. Alonso, M.S.; Fidalgo, M.R.; Labajo, J.L. The urban heat island in Salamanca (Spain) and its relationship to meteorological parameters. *Climate Res.* **2007**, *34*, 39–46. [CrossRef]

64. Peng, S.; Piao, S.; Ciais, P.; Friedlingstein, P.; Ottle, C.; Bréon, F.M.; Nan, H.; Zhou, L.; Myneni, R.B. Surface urban heat island across 419 global big cities. *Environ. Sci. Technol.* **2012**, *46*, 696–703. [CrossRef]

65. Bahi, H.; Rhinane, H.; Bensalmia, A.; Fehrenbach, U.; Scherer, D. Effects of urbanization and seasonal cycle on the surface urban heat island patterns in the coastal growing cities: A case study of Casablanca, Morocco. *Remote Sens.* **2016**, *8*, 829. [CrossRef]

Article

The Cooling Effect of Large-Scale Urban Parks on Surrounding Area Thermal Comfort

Farshid Aram [1], Ebrahim Solgi [2], Ester Higueras García [1], Amir Mosavi [3,4,5,*] and Annamária R. Várkonyi-Kóczy [3,6]

[1]   Escuela Técnica Superior de Arquitectura, Universidad Politécnica de Madrid-UPM, 28040 Madrid, Spain
[2]   School of Engineering and Built Environment, Griffith University, Gold Coast 4222, Australia
[3]   Institute of Automation, Kalman Kando Faculty of Electrical Engineering, Obuda University, 1034 Budapest, Hungary
[4]   Faculty of Health, The Queensland University of Technology, Brisbane, QLD 4059, Australia
[5]   School of Built the Environment, Oxford Brookes University, Oxford OX30BP, UK
[6]   Department of Mathematics and Informatics, J. Selye University, 94501 Komarno, Slovakia
*   Correspondence: amir.mosavi@kvk.uni-obuda.hu

Received: 1 September 2019; Accepted: 14 October 2019; Published: 15 October 2019

Abstract: This empirical study investigates large urban park cooling effects on the thermal comfort of occupants in the vicinity of the main central park, located in Madrid, Spain. Data were gathered during hot summer days, using mobile observations and a questionnaire. The results showed that the cooling effect of this urban park of 125 ha area at a distance of 150 m could reduce air temperatures by an average of 0.63 °C and 1.28 °C for distances of 380 m and 665 meters from the park. Moreover, the degree of the physiological equivalent temperature (PET) index at a distance of 150 meters from the park is on average 2 °C PET and 2.3 °C PET less compared to distances of 380 m and 665 m, respectively. Considering the distance from the park, the correlation between occupant perceived thermal comfort (PTC) and PET is inverse. That is, augmenting the distance from the park increases PET, while the extent of PTC reduces accordingly. The correlation between these two factors at the nearest and furthest distances from the park is meaningful (p-value < 0.05). The results also showed that large-scale urban parks generally play a significant part in creating a cognitive state of high-perceived thermal comfort spaces for residents.

Keywords: cooling effect; urban park; thermal comfort; physiological equivalent temperature; perceived thermal comfort; urban heat island; air temperature; sustainable cities; smart cities; urban health; global warming; urban green spaces; sustainable urban development; climate change mitigation and adaptation; urban resilience

---

1. Introduction

Due to climate change and the growing urbanization, the heat in cities is rising rapidly [1,2]. Today, heat has adversely affected urban life across the world [3], including urban areas of the Mediterranean climate [4,5]. The increase in temperature in urban areas, especially densely populated areas has given rise to the phenomenon of urban heat islands (UHI) [6,7], which can threaten the health and comfort of citizens [8,9]. The green urban spaces have been researched by numerous studies as an adaptive strategy to reduce the effect of urban heat and improve the health of citizens, by considering thermal comfort [10,11] as well as their socio-economic role [12]. Various studies have been conducted on the effects of various types of green infrastructures aiming at reducing urban heat and thermal discomfort [13,14], which include different scales and forms such as small local parks [15], large urban parks [16], urban gardens [17], green roofs [18,19], green walls [20,21] and street trees [22,23]. This study emphasizes the cooling effect of large urban parks.

The studies conducted on large urban parks have shown that such parks have a significant effect on air temperature reduction and UHI [24–26], which is especially noticeable during the summer [27,28]. However, it is noteworthy that the temperature drop does not only take place within the park but also surrounding areas [29]. The cooling effect of urban parks, referred to as a park cooling island (PCI) [30], is known to affect the surrounding environment depending on the area of the green space and quality of green coverage [31,32]. Two relevant indices to measure the cooling effect of urban parks are the cooling effect distance (CED) and cooling effect intensity (CEI) [33], which have been extensively used in various scales and climates [34–36].

Studies on the cooling effect of urban parks in different regions have shown that large scale urban parks with areas over 10 ha can have an average of 1 to 2 °C effect on surrounding areas that are an average of 350 meters away [33]. In general, air temperature reductions in urban parks are typically up to 0.5–4 °C and may even cause up to 5–7 °C reduction [37]. Despite the cooling effect of urban parks on their surroundings, such an outcome is not merely dependent on the traits of the green space [38]. The morphology of the surrounding area of parks, the sky view factor (SVF), the spatial configuration of the location, and the covered area, also affect the perception of the cooling effect of urban parks and thermal comfort [37,39].

Numerous studies have been carried out on the cooling effect of urban parks. However, research pertaining to the extent of this effect on thermal comfort perception from psychological and physical perspectives is limited. This gap necessitates examining the cooling effect of urban parks on the aforementioned parameters at different intervals from the park to determine factors that may be applied in sustainable urban development.

2. Park Cooling Effect and Thermal Comfort Perception

A set of thermal indices was investigated in two experimental (effective temp (ET), resultant temp (RT), humid operative temp (HOP), operative temp (OP) and wind chill index (WCI)) and analytical (index of thermal stress (ITS), heat stress index (HIS), effective temp (ET*), stand. effective temp (SET*), out. stand. eff. temp (OUT_SET), predicted mean vote (PMV), perceived temp (PT) and physiological equivalent temperature (PET)) groups [40–43]. The basis of these analytical indices was the energy balance (produced and wasted human energy). The main issue of thermal indices pertains to the average thermal comfort assessment and climate conditions of each area [44]. The result of various studies on validating other indices shows that examples such as SET and PET have a high correlation (89%) with thermal comfort in open spaces [45]. Most studies in recent years have utilized SET, PMV and PET to predict comfort levels in open spaces [46–48].

PET enables the comparison of the full effect of thermal conditions about the outside environment with individual experience [49,50]. PET is one of the recommended indices in urban and regional urban planning around the world, used to predict thermal changes of urban or regional clusters [51]. This index has shown a significant correlation to thermal comfort in various climatic conditions in open urban spaces as validation [52]. One of the prominent influencing factors on PET condition is the T_{mrt} (mean radiant temperature) climate variable. The indices mentioned above provide a single image of a set of individual and climate variables and enable the comparison of comfort conditions in various environments (due to global factors) [53].

In studies conducted to investigate the thermal comfort of urban parks, the PET index is commonly used. In a study conducted on a warm sunny day in Beijing, China (21 August, 2 pm) [54], it was shown that the effect of the Yuan Dynasty Relics green space with 102 ha area, in close approximately to the area of Retiro Park, reduces the PET by an average of 2 °C and a maximum of 15.6 °C. Another study conducted at Zhongshan Park in Shanghai city center with an area of 21.42 ha, showed that the PCI led to thermal comfort during Shanghai's winter as well as summer and had a PET of 15–29 °C [55]. Additionally, a study conducted at the Central Park of Cairo with an area of approximately 26.01 ha [56] showed that parks had a significant effect in enhancing thermal comfort during summer, which, based on this effect, entailed a PET value of 22–30 °C throughout the day.

While the cooling effect of urban parks has generally been recognized as a strategy for mitigating urban heat and providing thermal comfort [57,58], the extent of perceived thermal comfort stemming from the cooling effect of urban parks is difficult to predict, necessitating investigations of people's attitudinal and bodily experiences concerning physical thermal conditions at various intervals in the surrounding areas of these parks [59,60]. The degree of perceived thermal comfort by an occupant, which is caused by the cooling effect of urban parks, is dependent on factors such as individual behavioral and psychological traits, in addition to distance and cooling effect intensity [61–63]. Individual demographic traits include age, gender, and physical characteristics such as height and weight of occupants [64,65]. Psychological traits include the individual's experience of being in the environment and their thermal expectations and tolerance [66,67]. Behavioral aspects include the extent of being covered (in terms of clothes) and the type of activities conducted by the individual [68].

A common method to determine the extent of thermal comfort experienced by occupants is the use of surveys [69]. Essentially, by using such survey methods as open, semi-open, or multiple-choice questions [70], individuals can be asked about their thermal comfort experiences [71]. In order to measure the human thermal comfort perception level, cognitive mapping can be utilized in combination with questionnaires. Employing this method and asking residents to identify places where they feel more comfortable from a thermal point of view provided a more comprehensible image of the residents' thermal comfort level for understanding [72,73]. In recent years, utilizing questionnaires and asking direct questions about the perceived comfort of citizens alongside micro-climatic perceptions have yielded valuable results [74].

Many studies have been conducted on the cooling effect of urban parks, particularly their CEI and CED effects. However, less research has been carried out on the thermal comfort of this effect at different distances from the park. Additionally, few studies on the cooling effect of urban parks on thermal comfort have mainly investigated through physiological indices. As both psychological and physical perspectives together provide the complete concept of thermal comfort, it, therefore, necessitates examining the extent of thermal comfort created by urban green spaces from either perspective.

The purpose of this study is to investigate the cooling effect of the large urban park and its effect on the thermal comfort of occupants in the areas around the park. This study focuses on the perceived thermal comfort (PTC) and physiological equivalent temperature (PET) of locations, at a defined distance from the park and similar in terms of urban aspects and influential factors such as floor coverings, enclosures, street canyon and vegetation, whilst responding to the following research questions:

- What is the extent of the large urban park cooling effect on thermal comfort based on PET?
- What is the extent of the large urban park cooling effect on occupants' PTC?
- What is the relationship between the measured PTC and the measured occupant PET?

3. Materials and Methods

3.1. Location and Selection of the Sites

This research was located in Madrid (40°25′08″ N; 3°41′31″ O), the capital of Spain with a population of 3,223,334 and population density of 5265.91 /km^2, and a hot-summer Mediterranean climate (Csa) according to the Köppen classification [75]. The average annual temperature in downtown Madrid is 19.9 °C during the day and 10.1 °C at night. The warmest month of the year is July, with an average temperature of 32.1 °C during the day. Then August, June and September are the hottest months with average daily temperatures of 31.3 °C, 38.2 °C and 26.4 °C, respectively [76]. Retiro Park has an area of approximately 125 ha. It is the largest and oldest park in the center of Madrid with rich plant diversity. One of the prominent issues of this study was the method of selecting distances from Retiro Park. Although studies have not been conducted on the average cooling effect of Retiro Park, an updated UHI map of Madrid was presented in 2015 by Núñez Peiró et al. [77], where the effect of Retiro Park was made evident based on the temperature color spectrum. Essentially, the UHI map of

Madrid from 26 July 2015 represents the effect of Retiro Park in counteracting to the effect of UHI in the northern, eastern and southern areas of the park [78].

In order to accurately assess the cooling effect of Retiro Park on its surrounding areas, it was necessary to select places on a micro-scale with common features that are located at various distances from the park. The northern area of Retiro Park, due to its well-organized form, has more regular urban-type compared to other areas (east and south). Based on the temperature zones of the updated thermal map of Madrid's urban heat islands in 2015 [79], three street crossings (intersections) located in the Salamanca and Recoletos neighborhoods in the north of Retiro Park along the Lagasca Street, were chosen. These sites were located at 150 meters, 380 meters and 665 meters from the park at Lagasca intersection with the Conde de Aranda, Jorge Juan and Hermosilla streets, giving zones of different temperature ranges (yellow, orange and red; Figure 1). These types of selected local intersections represent most of Retiro Park's northern area intersections (Figure 2). Based on this map [77,79], each of the three intersections A, B and C had a temperature difference of about 0.8 °C.

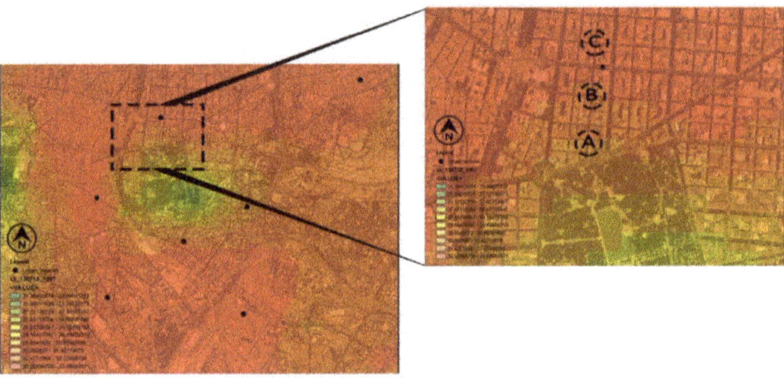

Figure 1. Madrid's urban heat island (UHI) map on 26 July 2015 [77,79] and the selected northern region of the Retiro Park. Intersection A is in the yellow zone, intersection B is in the orange zone and intersection C is in the red zone, close to one of the UHIs of Madrid.

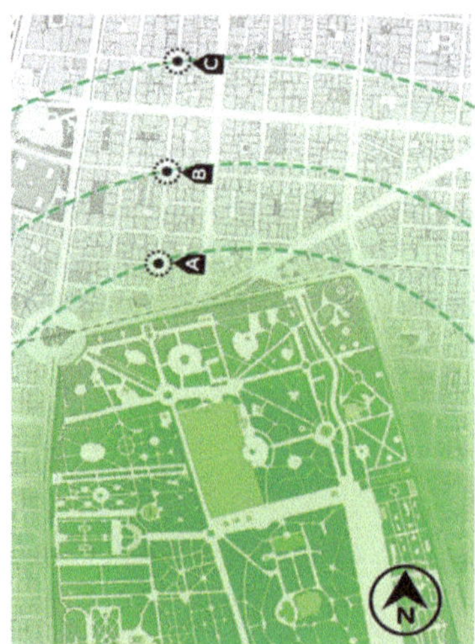

Figure 2. The northern region map of Retiro Park and the location of the selected sites. Point A is located 150 meters away from the park; point B is located 380 meters away from the park and point C is located 665 m away from the park.

All three assessed intersections were selected based on a common geometric configuration. The considered geometric configuration included an enclosure, sky view factor (SVF), and street height and width (Figure 3, Table 1). The considered material structure included the ground surface coverage material, pavement material and building facade material, which were similar in all three regions; the ground surface material in all three regions was asphalt; the pavement material was gray tiles and the building material is mostly red-colored bricks and bright cement (Figure 3). The presence of green spaces in the street canyon was highly significant in terms of micro-climatic issues [80]. In this regard, vegetation was also considered as well as the aforementioned parameters (Table 1), so the selected regions were also similar in terms of this parameter (Figure 3).

Figure 3. Street view of the three intersections (**A–C**) as well as fisheye shots.

Table 1. The geometric configuration of the intersections and street trees properties in three investigated intersections.

Part	Street Canyons					Street Trees			
	Width N_S (M)	Width W_E (M)	H/W Ratio N_S (–)	H/W Ratio W_E (–)	SVF [a]	Number of trees	Mean Tree Height (m)	Mean Crown Diameter (m)	Crown Shape [b]
A	15	12	1.23	1.54	0.17	26	12.74	5	Cone
B	15	15	1.28	1.33	0.2	30	12.34	6	Cone
C	15	15	1.2	1.23	0.2	25	12.66	5	Cone

a. Calculated by Ray Man 1.2. b. Crown Shape Classification by Park et al. [15].

There were two priorities in selecting the intersections. Firstly, the points were selected according to Madrid's UHI map (Figure 1) where the distances were of different temperature ranges, and secondly, locations were selected that shared the greatest similarity in terms of physical and structural traits. In this vein, three sites A, B and C were selected to extract data, in accordance with the goals of this study.

3.2. Microclimate Parameter Measurements

Micro-climatic measurements were conducted during six days in hot summer, on 22 June, 10 and 24 of July, 10 and 24 August and 10 September 2018. Data extraction started from 22 June and was spaced at roughly every two weeks, targeting sunny and calm days (no clouds) up to the end of summer (10 September). Ta and relative humidity (RH) were collected during the six days, were measured by mobile microclimate stations (HOBO MX2301A Temperature/RH Data Logger, manufactured by Onset Computer Corporation, MA, USA) with precision T_a; ± 0.2 °C and RH; ± 2.5% between 10:40 and 12:10 CEST (Central European Summer Time). The data was collected at one-minute intervals, and the weather data collection duration at each location was 10 minutes (Table 2).

Table 2. Intersection mean values for air temperature (T_a), relative humidity (RH) and wind speed (WS) on all measurement days (10:40–12:10 CEST).

Date	Part	Time	Mean T_a, °C	Mean RH %	Mean WS, m/s
	C	10:40–10:50	29.26	29.1	1.43
22 June	B	11:00–11:10	28.87	29.27	1.15
	A	11:30–11:40	27.81	32.04	0.80
	C	11:15–11:25	29.8	33.96	1.43
10 July	B	11:42–11:52	29.7	22.1	1.16
	A	12:00–12:10	29.4	29.71	1.36
	C	10:40–10:50	29.45	21.71	1.36
24 July	B	11:10–11:20	28.1	30.25	1.43
	A	11:30–11:40	27.74	30.56	1.43
	C	10:40–10:50	25.12	18.16	1.43
10 August	B	11:00–11:10	23.77	41.05	1.72
	A	11:20–11:30	23.22	43.45	2.47
	C	10:50–11:00	26.63	38.81	2.35
24 August	B	11:15–11:25	26.2	38.13	1.13
	A	11:35–11:45	26.05	37	1.29
	C	10:45–10:55	23.41	56.71	1.5
10 September	B	11:15–11:25	23.13	52.45	1.36
	A	11:35–11:45	21.73	51.15	1.91

The devices were placed 1.5 meters from the ground and covered by a radiation shield. The wind speed (WS) was determined by a Proster Digital Anemometer MS6252a placed at the same distance. The type of ground coverings, walls and types of and distance to vegetation that affect the temperature conditions were checked according to field studies; high-resolution images were taken by a Canon Eos 600D, 5184 pixels × 3456 pixels digital camera. Arial images were taken from Google Earth, 2018–2019. Fisheye images were taken at three intersections using a fisheye (Sigma 8 mm circular) lens. Additionally, the weather data of Retiro Park were collected via the AEMET (Agencia Estatal de Meteorologia) fixed station [76] located inside the park. These parameters were previously identified to show the effect of urban parks on thermal comfort [54].

3.3. The Questionnaire Survey

A random semi-structured survey was conducted on 145 pedestrians (nA: 49, nB: 45 and nC: 51) at intersections where microclimatic data were collected (workplace and residential). The interviewees comprised different gender and age groups (with the exception of children), and were active at different levels. Table 3 shows detailed information on the number and characteristics of individuals, on different days and intersections.

Data extraction from the surveys took place during the period of determining micro-climatic parameters at three intersections A, B and C (more and less than 10 minutes). Questions were divided into three sections.

The second section comprised four questions about the degree of thermal comfort perception during the interview. The questions were short and designed as five options (very high = 5 to very low = 1). Research questions concerning the level of individual thermal comfort included: (Q1) How much thermal comfort do you feel? (not too hot, not too cold); (Q2) How warm do you feel? (Q3) What is the extent of thermal comfort perceived through Retiro Park? (Q4) How much heat can you tolerate in this location?

The third section included open personal questions such as gender, height, weight, level of activity and type of clothing.

Table 3. The proportional percentages questionnaire data in the three investigated intersections on all the measured days.

Variable	Categories	Percentage (%)		
		A	B	C
Age	13–21	2.0	2.2	11.8
	22–30	26.5	20.0	13.7
	31–45	40.8	26.7	31.4
	46–60	16.3	24.4	29.4
	61–85	14.3	26.7	13.7
Gender	Men	67.3	73.3	51
	Women	32.7	26.7	49
Q1	Very low (1)	0	0	0
	Low (2)	10.2	6.7	9.8
	Medium (3)	28.6	57.8	43.1
	High (4)	42.9	31.1	35.3
	Very high (5)	18.4	4.4	11.8
Q2	Very low (5)	5.9	4.4	0
	Low (4)	23.5	31.1	18.4
	Medium (3)	33.3	26.7	22.4
	High (2)	27.5	28.9	38.8
	Very high (1)	9.8	8.9	20.4
Q3	Very low (1)	4.1	2.2	13.7
	Low (2)	8.2	20.0	25.5
	Medium (3)	28.6	31.1	27.5
	High (4)	42.9	31.1	25.5
	Very high (5)	16.3	15.6	7.8
Q4	Very low (1)	2.0	0	3.9
	Low (2)	10.2	15.6	25.5
	Medium (3)	30.6	60.0	35.3
	High (4)	42.9	13.3	23.5
	Very high (5)	2.0	0	3.9

Essentially, the questionnaire questions were designed so that in a short period (10 minutes), pedestrians in the neighborhood of different age and gender groups could be interviewed to extract information about their level of perceived thermal comfort (Figure 4). The total score from the questions is presented as an indicator of the perceived thermal comfort (PTC). However, the inquiry of PTC was ambiguous for occupants, so this question was addressed by asking how comfortable do you feel in terms of temperature (not too hot not too cold)? For the second inquiry, the question that was also directly asked was how much heat do you feel? Therefore, data relevant to this question was considered in reverse form to determine the perceived thermal comfort index (very high = 1 to very low = 5). It should be noted that in order to accurately derive the research data, prior to the start of summer on 22 May, 18 experimental micro-climatic data and questionnaires (six questionnaires at each intersection) were compiled (10 minutes at each intersection) in order to address the issues and queries of questionnaire gaps and to identify bio-climatic notions amidst the main questionnaire compilation stage.

3.4. Analyzing the Thermal Comfort Parameters

3.4.1. Physiological Parameters

In order to analyze thermal comfort and derive PET, the relevant parameters should be calculated. The calculation of parameters for PET assessment includes clo, level of activity, SVF and T_{mrt}. In this study, Ray Man 1.2 [81,82] software was used to derive SVF, T_{mrt} and PET. The calculation of clo, which pertains to the extent of being covered for an individual, was based on its computational indices [83]. Given that each clothing item has its own index, the clo for each person was based on the collected data. The greatest extent of clothing coverage for participants was a T-shirt with trousers (34%), shirt with trousers (25.4%) and the least was shorts with a t-shirt (22.6%), which were of 61 clo, 65 clo and 40 clo, respectively. Determination of activity levels was based on individual activity in the environment.

Every activity had an indicator, walking 115 W/m^2, sitting 60 W/m^2, standing 70 W/m^2, and fast walking 220 W/m^2 [84] (Table 4).

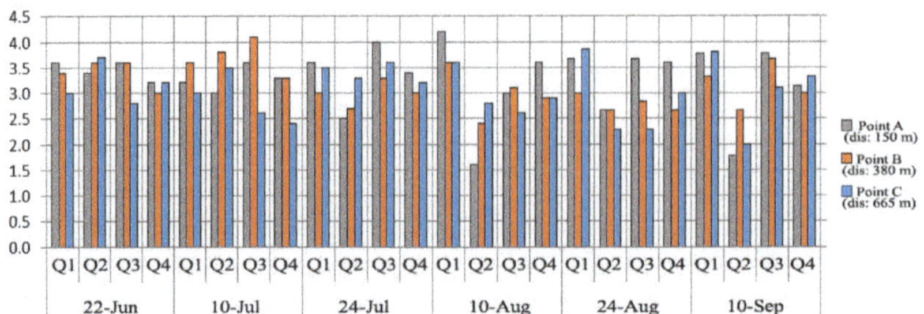

Figure 4. The average frequency of questionnaire data (Questions 1–4) at three intersections. Intersection A is gray, intersection B is orange and intersection C is blue.

Table 4. Clothing and activity level of responders in the three investigated intersections.

Variable	Percentage (%)																	
	Hot Summer Days																	
	22 June			10 July			24 July			10 August			24 August			10 September		
	A1	A2	A3	A1	A2	A3	A1	A2	A3	A1	A2	A3	A1	A2	A3	A1	A2	A3
_Clothing																		
Shirt and Normal Pants - 65 clo	40	40	33.3	44.4	33.3	25	30	33.3	10	0	20	0	0	16.6	28.5	22.2	50	30
Tshirt and normal pants - 61 clo	20	20	33.3	22.2	22.2	37.5	30	44.4	50	30	40	50	33.3	16.6	57.1	33.3	33.3	50
shirt and Shorts - 45 clo	0	0	0	11.1	0	0	10	0	0	10	10	0	16.6	16.6	0	11.1	0	10
Tshirt and Shorts (or skirt) - 40 clo	40	20	16.6	11.1	33.3	0	20	22.2	20	50	10	40	50	33.3	14.3	0	16.7	10
Dress - 35 clo	0	0	0	11.1	11.1	12.5	10	0	10	10	10	0	0	0	0	0	0	0
Suiet - 90 clo	0	0	16.6	0	0	12.5	0	0	10	0	10	10	0	16.6	0	11.1	0	0
Others	0	20	0	0	0	12.5	0	0	0	10	0	0	0	0	0	22.2	0	0
_Activity																		
Wlking - 115 W/m^2	60	10	66.6	55.5	77.8	50	40	44.4	60	30	50	80	50	66.6	57.1	55.5	83.3	40
Standing - 70 W/m^2	20	0	16.6	44.4	22.2	12.5	50	44.4	20	40	30	10	16.6	33.3	14.3	11.1	16.7	20
Sitting - 60 W/m^2	20	0	16.6	0	0	37.5	10	0	20	30	20	10	33.3	0	28.5	33.3	0	40
Jogging - 220 W/m^2	0	0	0	0	0	0	0	11.1	0	0	0	0	0	0		0	0	0

SVF calculation was conducted by importing fisheye images in the RayMan software, and 3D simulation of spatial geometry and environment vegetation in the Obstacle section of the software. The SVF index is one of the effective parameters in deriving the thermal comfort of the environment [85].

The PET index was considered a vital indicator in studying thermal comfort and has been used in numerous studies as a reliable indicator in deriving the thermal comfort of the external environment [46]. T_{mrt} and PET were derived based on extracted micro-climatic data (Table 5), and SVF data derived via the Ray Man software (Table 1).

3.4.2. Psychological Parameters

SPSS software was utilized, according to the statistical data obtained from the questionnaire to examine the significance of the data in three selected intersections, and also to examine the significance of the relationships between various indices with PTC [86,87]. In order to examine cognitive maps more accurately, AramMMA software was used [88]. In this program, all the cognitive maps were overlapped, and then according to the color spectrum, it determined which locations had the most point of reference (Figure 5). Each color represents the number of pointing a location (for example red means one time or purple means 12 times). In order to convert the qualitative data of cognitive maps into quantitative data, inside the AramMMA software, the Retiro area was rated, so that 100 percent of the score was dedicated to the maps that noted parks, while zero percent was considered for those that did not mention to the park. By using the cognitive map analyses, the importance of park cooling effect on resident psychological thermal comfort in different parts of the Retiro Park was represented.

Table 5. Values of the T_{mrt} (°C) and average values for physiological equivalent temperature (PET; °C) in part A, B and C.

Date	Part	Mean PET, °C	T_{mrt}, °C
	C	35.3	48.4
22 June	B	35.8	48.7
	A	33.7	48
	C	36.3	49.4
10 July	B	36.5	49.3
	A	35.8	49.3
	C	35.7	49
24 July	B	33.9	47.7
	A	33.5	47.6
	C	28.9	43.8
10 August	B	26.8	43.1
	A	24.3	42.7
	C	29.1	44.4
24 August	B	28.6	44.5
	A	26.9	44.3
	C	26.1	40.6
10 September	B	26.2	41
	A	23.1	39.8

Figure 5. Cognitive map analyses using the AramMMA software [88] during six hot summer days of 2018.

4. Results

Data was gathered during six days in hot summer, on 22 June, 10 and 24 July, 10 and 24 August and 10 September at three different intersections at different distances from the northern area of Retiro Park (Table 2). The selected data collection time interval was generally between 10:40 and 12:10 CEST (the exact time of each intersection is given in Table 2). This time was chosen since it was between the maximum and minimum temperatures in the park, so the temperature data extracted at points A, B and C were close to the average park temperature (mid-temperature taken by the AEMET weather station [76] on the data extraction days; Table 6). Based on Figure 6, the temperature range for points A, B and C (gray rectangle) was between the maximum and minimum extracted park temperature (linear range). It is noteworthy that among the three areas, the average temperature of point A was closer to the average temperature of the park (mid-temperature; Table 6) compared to the other points. AEMET data showed that the lowest and highest temperatures pertaining to the data extraction days in the Retiro Park were related to the 4:50–6:00 and 13:40–14:40 CEST timeframes.

Table 6. Mean values for air temperature (T_a), in the three intersection (10:40–12:10 CET) and values for air temperature (T_a), relative humidity (RH) and wind velocity (W) in the Retiro Park on all the measurement days.

Date	T_a [a] (°C)			Retiro Park [b] T_a (°C)			Time of T_a in Retiro Park [b]		HR % Park [b]	Wind in Park [b]
	Part A	Part B	Part C	Min	Mid	Max	Min	Max		
22 June 2018	27.81	28.87	29.26	21.6	27.7	33.8	04:50	14:40	22.95	1.7
10 July 2018	29.4	29.7	29.8	21.5	28.4	35.2	06:00	13:50	22.95	2.2
24 July 2018	27.74	28.1	29.45	19.8	26.4	33.0	05:00	13:50	22.94	1.9
10 August 2018	23.22	23.77	25.12	17.5	24.4	31.3	05:40	13:40	36.8	2.2
24 August 2018	26.05	26.2	26.63	20.6	26.8	33.0	05:30	14:20	19.25	1.4
10 September 2018	21.73	23.13	23.41	17.3	22.8	28.3	05:20	13:45	33.55	1.9

a. Field measurement, b. AEMET data.

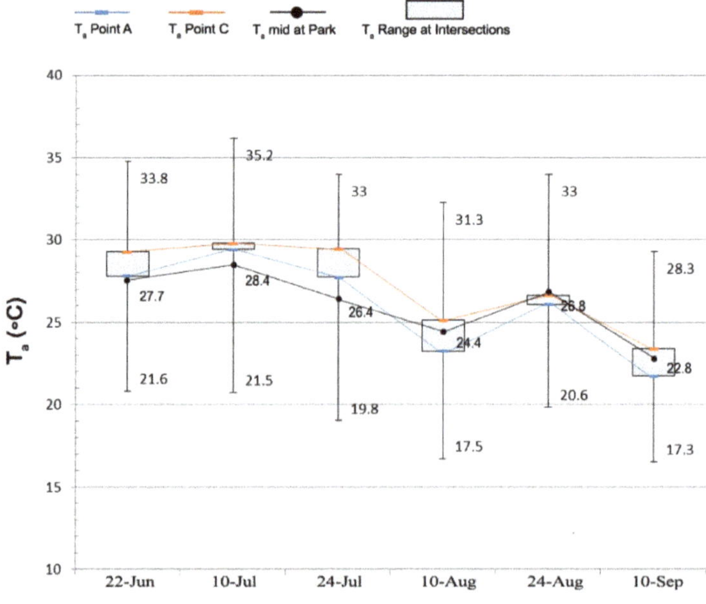

Figure 6. The diagram shows the temperature range of the three intersections A, B and C measured and the temperature range of Rétiro Park during the six days of hot summer 2018 in Madrid.

Gray rectangles: Temperature range at three intersections (bottom side: T_a A, top Side: T_a C).

Linear range: The temperature range of the park is based on AEMET (bottom line: T_a mid of park, top line: T_a Max of the park.

The mean T_a for points A, B and C are 25.99 °C, 26.62 °C and 27.27 °C, respectively. The T_a difference over the measurement days between distance of 150 m (A) and 380 m (B) from the park was about 0.63 °C, between B and C (distance of 665 m) was about 0.65 °C and for the difference between A and C, the temperature difference was about 1.28 °C. According to the results, as the distance from the park increases, the temperature and its difference with areas closer to the park will be greater.

4.1. Assessment of the Park's Cooling Effecting on Thermal Comfort Indices

In this section, PET data and questionnaire results were used to assess the cooling effect of Retiro Park on thermal comfort. PET data are derived based on the indices standard and the perceived thermal effect is obtained based on the average questionnaire and cognitive maps data. By using both these data, the extent of thermal comfort originating from the Retiro Park cooling effect at distances A, B and C could be deduced as physiological and psychological points of view.

4.1.1. Comparison of the Park Cooling Effect Significance on PET at Different Distances from the Park

In a more accurate analysis using SPSS software, a one-way ANOVA was used to analyze the relationship between distances A, B and C from the park and the relevant PET differences. Upon every variance analysis and in the case of significant mean difference (the significant level of a p-value less than 0.05), Tukey test, which is a series of post-hoc tests, were used to accurately determine which of the average of a variable has a significant difference [89]. Essentially, by using this test, the thermal comfort relationship was deduced based on the PET index, which was obtained using the Ray Man software on distances A, B and C.

The results of the ANOVA test (Table 7) show that concerning the PET index, there was a lower significant error rate (0.05) and 95% confidence level of a significant difference between average PET data in each of the A, B and C points (p-value = 0.029). Therefore, Tukey's post-hoc tests were used to determine whether there is a significant difference between each of the A, B and C intersections.

Table 7. ANOVA PET analyses in the three investigated selected points.

	Sum of Squares	df	Mean Square	F	p-Value
Between Groups	151.731	2	75.865	3.645	0.029
Within Groups	2955.708	142	20.815		
Total	3107.439	144			

The results of this studied grouping test are presented in Table 8. According to this table, it can be stated that there was a significant relationship between the mean PET index at points A and C but the mean PET index between point B with points A and C did not entail a significant difference. The mean PET for A, B and C was 29.3 °C, 31.3 °C and 31.6 °C, respectively. Essentially, it can be stated that the PET index difference among intersections A and C was explicit and significant (p-value = 0.036). The mean of this index in A had a difference of 2.3 °C with C, while the difference between A with B was about 2 °C, which was not considered significant in terms of the Tukey test (p-value = 0.09).

4.1.2. Comparison of the Significance of the Questionnaire Dataset PTC at Different Distances from the Park

In order to compare the significance of the PTC at different distances from the park, the mean dataset extracted from the questionnaire for each section was derived as the PTC. The questions answered by citizens via the questionnaire were about how they felt about the ambient temperature and the effect of Retiro Park on such feelings in the form of four questions (Table 3).

Table 8. Tukey PET analyses in the three investigated intersections (A, B and C).

(I) Part	(J) Part	Mean Difference (I–J)	Std. Error	*p*-Value	95% Confidence Interval		Part	*N*	Subset for alpha = 0.05	
					Lower Bound	Upper Bound			1	2
A	B	−1.99624	0.94199	0.090	−4.2274	0.2349	A	49	29.3327	
	C	−2.28303 *	0.91265	0.036	−4.4447	−0.1214	B	45	31.3289	31.3289
B	A	1.99624	0.94199	0.090	−0.2349	4.2274	C	51		31.6157
	C	−0.28680	0.93311	0.949	−2.4969	1.9233				
C	A	2.28303*	0.91265	0.036	0.1214	4.4447	Sig.		0.084	0.949
	B	0.28680	0.93311	0.949	−1.9233	2.4969				

*. The mean difference is significant at the 0.05 level.

Based on the results of the ANOVA statistical test (Table 9) and a lower significance error level (*p*-value < 0.05), there was 95% confidence pertaining to a significant difference between mean thermal comfort index (PTC) in each of the A, B and C regions (*p*-value = 0.032). Thus, Tukey's post-hoc test was used to determine the significant difference between each of the intersections.

Table 9. ANOVA perceived thermal comfort (PTC) analyses in the three investigated selected points.

	Sum of Squares	df	Mean Square	F	*p*-Value
Between Groups	2.318	2	1.159	3.517	0.032
Within Groups	46.809	142	0.330		
Total	49.128	144			

Since the rating of questions was from very low to very high in the form of numbers 1 to 5, Table 10 shows that the average of the total questions during the 6 days for A, B and C was approximately 3.3., 2.9 and 2.8, such that area A was of the highest share indicating more thermal comfort experienced by people in this area. In interpreting the table above, it can be stated that there was a significant difference between the mean PTC index at points A and C (*p*-value < 0.05), but the mean PTC at point B with points A and C had no significant difference. Based on this relationship, it is also clear that with an increase in distance from the green area, the perceived comfort experienced by citizens was reduced; thus there was an inverse correlation between increased distance and PTC.

Table 10. Tukey PTC analyses in the three investigated intersections (A, B and C).

(I) Part	(J) Part	Mean Difference (I–J)	Std. Error	*p*-Value	95% Confidence Interval		Part	*N*	Subset for alpha = 0.05	
					Lower Bound	Upper Bound			1	2
A	B	0.35417	0.14270	0.062	−0.0165	0.7248	C	51	2.8417	
	C	0.45000 *	0.14270	0.017	0.0793	0.8207	B	45	2.9375	2.9375
B	A	−0.35417	0.14270	0.062	−0.7248	0.0165	A	49		3.2917
	C	0.09583	0.14270	0.783	−0.2748	0.4665				
C	A	−0.45000*	0.14270	0.017	−0.8207	−0.0793	Sig.		0.783	0.062
	B	−0.09583	0.14270	0.783	−0.4665	0.2748				

*. The mean difference is significant at the 0.05 level.

4.1.3. Comparing the Cognitive Maps Significance at Different Distances from the Park

The meaningfulness of the cognitive maps data with the distance from the park was evaluated using the ANOVA test. The analysis of this section was performed by using a AramMMA software for analyzing cognitive maps. In this analysis, the maps pointing to the Retiro park were distinguished from maps that did not mention the park. In fact, 100 percent of the score was dedicated to the maps noting parks, while zero percent was considered for those that did not mention to the park. The result of the ANOVA test (Table 11) illustrated that there was an acceptable agreement (95%) between the correlation of cognitive maps and the distance from the park (*p*-value < 0.05).

Table 11. ANOVA cognitive maps analyses in the three investigated selected points.

	Sum of Squares	df	Mean Square	F	*p*-Value
Between Groups	36,309.751	2	18,154.876	8.948	0.000
Within Groups	288,104.042	142	2028.902		
Total	324,413.793	144			

Owing to the agreement, in each of the A, B and C regions the Tukey test was used separately, and the results were as follows. As illustrated (Table 12), 87.7%, 60% and 50.9% of the respondents in the A, B and C areas, respectively, referred to the park. As expected, the highest level of cognitive mapping from the park is in the vicinity of the park area (150 to 380 meters). Nevertheless, in the C region, with 665 meters distance from the park, more than half of the respondents claimed that the site was comfortable thermally thus preferring to spend more time there. The difference between A and C was also meaningful (*p*-value = 0.000). The good agreement of this relationship depicts the impact of the Retiro park (as a place with a high level of thermal comfort) on the resident's mental images and psychological dimensions.

Table 12. Tukey cognitive maps analyses in the three investigated intersections (A, B and C).

(I) Part	(J) Part	Mean Difference (I–J)	Std. Error	*p*-Value	95% Confidence Interval		Part	N	Subset for alpha = 0.05	
					Lower Bound	Upper Bound			1	2
A	B	27.75510 *	9.30015	0.009	5.7273	49.7829	C	51	50.9804	
	C	36.77471 *	9.01047	0.000	15.4330	58.1164	B	45	60.0000	
B	A	−27.75510 *	9.30015	0.009	−49.7829	−5.7273	A	49		87.7551
	C	9.01961	9.21244	0.591	−12.8005	30.8397				
C	A	−36.77471 *	9.01047	0.000	−58.1164	−15.4330	Sig.		0.589	1.000
	B	−9.01961	9.21244	0.591	−30.8397	12.8005				

*. The mean difference is significant at the 0.05 level.

4.1.4. Comparison of the Significance of the Questionnaire Dataset PTC with PET

For a closer assessment of the effect of Retiro Park on thermal comfort, the PET index and PTC by citizens were compared on the basis of the total data extracted from the questionnaire. Mainly, this PET analysis, which is a standard index for perceived thermal comfort, was compared with the personal opinion of people about their perceptions concerning the thermal comfort of the environment. SPSS software was used to assess the correlation test as well as other statistical tests.

Based on the results of the Pearson correlation test (Table 13), the relationship between the PET index with thermal comfort in the selected points of A and C was significant (*p*-value < 0.05) and was of inverse correlation (−0.404 and −0.379). Due to the medium correlation coefficient (medium = −0.3 to −0.5) [90–92], the relationship between the PET index and the PTC in area A (*p*-value = 0.004) was more significant that of area C (*p*-value = 0.006). Moreover, the significance of PET index relationship with perceived thermal comfort in area B (*p*-value = 0.061) was not accepted (*p*-value > 0.05).

Table 13. Pearson correlation analyses between PET and PTC.

		Part	PET
A	PTC	Pearson Correlation	−0.404 *
		p-value (2-tailed)	0.004
		N	49
B	PTC	Pearson Correlation	−0.281
		p-value (2-tailed)	0.061
		N	45
C	PTC	Pearson Correlation	−0.379 *
		p-value (2-tailed)	0.006
		N	51

*. The mean difference is significant at the 0.05 level.

4.2. Relations Between Thermal Comfort Indices (Physiological and Psychological)

Regarding the results, by increasing the distance from the park at intersections B and C, despite physical and structural similarities to the intersection the A, there was an increase in air temperature (T_a), and consequently, the extent of PET and T_{mrt} also changed accordingly.

The data pertaining to the perceived thermal comfort (PTC) acquired from the questionnaire was rated from 1 to 5 (1 = very low to 5 = very high) and at points A, B and C during six days were on average approximately 3.3, 2.9 and 2.8, respectively (Table 10). At point A, residents experienced higher thermal comfort compared to the other points. Essentially, the level of PTC at this location was medium to high according to citizens' opinions (medium = 3), whereas for B and C, the average perceived thermal comfort for the data extraction days was lower than medium. Furthermore, the results show that the extent of PTC at intersection A was more significant in all the days (Figure 7). It is noteworthy that based on statistical data, the perceived thermal comfort was correlated to PET and was of inverse correlation, such that its graph behavior was inversely correlated with the PET graph (Figures 7 and 8). For example, on 10th August, where the PET value was at a minimum at point A (PET 24.3 °C), the relevant perceived thermal comfort was at a maximum (3.6) or on 10th July, at point C, the PET was 36.3 °C, which was higher compared to other points and days which was inversed in the perceived thermal comfort graph and the least value of this index was at point C with a value of 2.4. In addition to the similar behaviors of PET, T_{mrt}, T_a and PTC parameters at each intersection during the six days of assessment (Figures 7 and 8), the credibility of the data and calculations conducted in this study were approved.

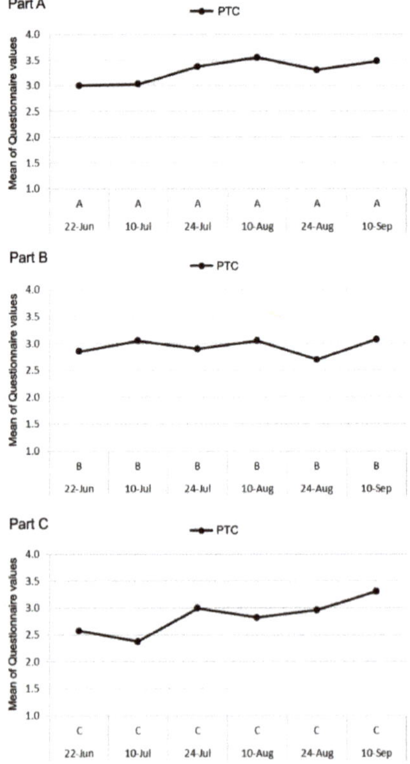

Figure 7. The average data for questionnaire data titled PTC at each three intersections A, B and C during six days.

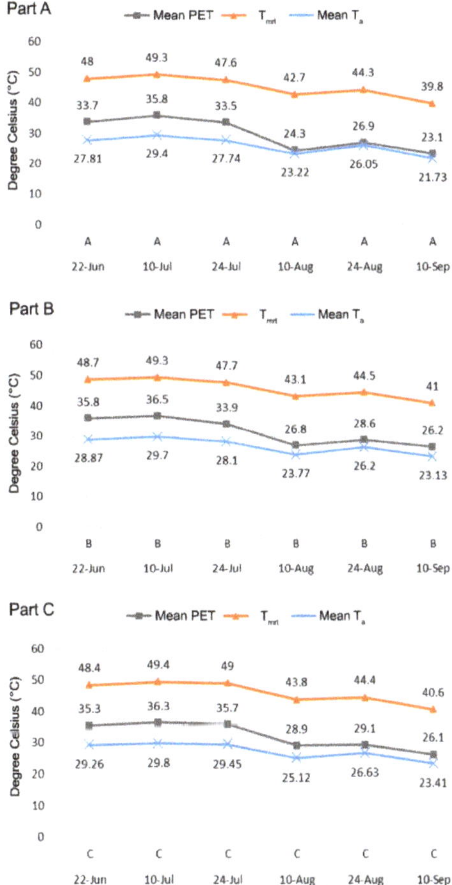

Figure 8. The relationship among mean PET (gray), mean T_a (blue) and T_{mrt} (orange), at each three intersections A, B and C during six days.

Based on the results, it can be said that Retiro Park played an essential role in providing thermal comfort to the citizens during the summer days in downtown Madrid, which was physically and psychologically debatable. As mentioned, Madrid has a Mediterranean climate that is characterized by hot summers. Studies in the Mediterranean climate have shown how urban green spaces can reduce the impact of urban heat [93,94]. A study carried out in Greece, Athens [95], found that the urban park on the western margin of Athens during the hot summer days could reduce daytime air temperature between 0.2 °C and 2.6 °C. In a similar study conducted in Lisbon, another Mediterranean city during the six days of summer 2007 and 2008 [96], the results showed that a 0.24 ha urban park was able to reduce the air temperature inside the park to 6.9 °C compared to its surrounding air temperatures. These studies and the studies mentioned in the first part of the article indicate that large-scale parks affect their surroundings through cooling down their environment. This is significant for Retiro Park due to its variety of vegetation and vegetation density compared to previous studies. The cooling effect of the park can cool over a distance of 350 m in its northern part, with a dense and regular texture, between 0.06 and 1.28 °C compared to the 655 m range near Heat Island.

As noted earlier, little is known about the cooling effect of urban parks on thermal comfort, and most of the studies have focused on the CED and CEI levels. However, in the Mediterranean areas, a study was carried out in Israel investigating the thermal comfort of urban parks using the PET index.

In this study conducted by Cohen, Potchter and Matzarakis [97], a total of 10 urban parks of various areas (0.2 to 0.36 ha) were assessed in Tel Aviv. Although the investigated parks were smaller in terms of area compared to Retiro Park, the results illustrated that parks with richer vegetation density had greater cooling effects and thermal comfort, and could reduce temperatures by up to 3.8 °C whilst bringing PET to 18 °C during hot summer.

This study proved how important this is in areas with hot summers. However, in this study and other studies on the thermal comfort of large urban parks, less attention was paid to the thermal comfort from a psychological point of view. In this study, in addition to assessing CED and CEI of a park and thermal comfort from the physiological point of view (PET), mental thermal comfort through questionnaires (PTC) and cognitive maps were also investigated. Based on the results, the significance of both PTC and PET indices was confirmed. The two indices were inversely correlated, and when the cooling effect was reduced by the distance from the park, the PET rate increased and the PTC level decreased, which indicates the impact of the cooling effect of the large urban park on the thermal comfort from both psychological and physical perspectives.

In this regard, the role of large urban parks was tangible in providing thermal comfort for citizens and it is necessary to pay more attention to various levels of urban planning and sustainable development.

5. Conclusions

This study examined the potential of large urban parks in providing thermal comfort for citizens living within the park perimeter. The results extracted amidst six hot summer days in Madrid show that large urban parks exhibit a cooling effect. Considering the significance of the mean T_a difference between the distance of 150 m and 665 m from the park (p-value < 0/05), as well as the lower temperature of about 1.28 °C pertaining to the distance closest to the park compared to distances further away from the park (under equal conditions), in addition to the insignificant mean T_a difference at the distance of 380 m compared to 150 m and 665 m (p-value > 0/05), we found that the cooling effect of the large urban park at distances close to the park (under 380 meters) had a significant role in temperature reduction.

Accordingly, the level of PET would increase as the distance from the park increased, and residents would perceive less thermal comfort compared to distances closer to the park. The degree of PET index at a distance of 150 meters from the park was on average 2 °C PET and 2.3 °C PET less compared to distances of 380 meters and 665 meters respectively. The PTC of citizens was acquired based on the average obtained data from the questionnaire, which showed that people in the vicinity of the park experienced more thermal comfort. For the other two regions of the park, which were more distant, less thermal comfort was experienced (less than average).

The results of the cognitive maps analyses demonstrated that large parks played a significant role in thermal comfort improvement affecting people's mental map. For people, such parks are a space where they feel more comfortable, thereby spending more time to enjoy the desirable temperature. Although the resultant mental maps were closer to residents in the vicinity of parks, the results demonstrated that at long distances to the park (665 meters) those locations still had a psychological dimension in offering thermal comfort (more than 50%).

PET and PTC as the main variables of this research were inversely correlated such that when the distance from the park was increased, the PET was increased and thermal comfort was decreased. The correlation between the two indices was significant in the nearest and furthest distance from the park (p-value < 0.05) and the highest correlation coefficient of the two indices pertains to the distance closest to the park (150 meters). Essentially, this indicates the high level of perceived thermal comfort from the citizens' point of view as well as the PET compared to distant locations.

As a result, the cooling effect of a large urban park was a suitable solution to improve people's thermal comfort (as physiological and psychological perspectives) either within the park area or in the vicinity of the park. The results of this study clarified that in areas close to the park, such an effect on thermal comfort is more perceptible. Thus, in order to enhance the cooling effect of large urban parks, it is necessary to implement urban design and planning solutions (qualitative and quantitative) to provide more favorable conditions for the lives of citizens.

Author Contributions: Conceptualization, F.A. and E.H.G.; methodology, F.A. and E.S; software, F.A., A.M. and S.S.; validation, F.A., E.S. and E.H.G.; formal analysis, F.A. and A.S.; investigation, F.A.; data curation, A.M. and S.S.; writing—original draft preparation, F.A. and E.S; writing—review and editing, E.H.G., A.M. and A.R.V.K.; visualization, F.A. and A.R.V.K.; supervision, E.H.G and A.R.V.K.

Funding: This publication was supported by the project: "Support of research and development activities of the J. Selye University in the field of Digital Slovakia and creative industry" of the Research & Innovation Operational Programme (ITMS code: NFP313010T504) co-funded by the European Regional Development Fund. Further, the research has been co-funded by the European Regional Development Fund.

Conflicts of Interest: The authors declare no conflict of interest.

References

1. Dimoudi, A.; Kantzioura, A.; Zoras, S.; Pallas, C.; Kosmopoulos, P. Investigation of urban microclimate parameters in an urban center. *Energy Build.* **2013**, *64*, 1–9. [CrossRef]
2. Ninikas, K.; Hytiris, N.; Emmanuel, R.; Aaen, B. The Performance of an ASHP System Using Waste Air to Recover Heat Energy in a Subway System. *Clean Technol.* **2019**, *1*, 154–163. [CrossRef]
3. IPCC. *Meeting Report of the Intergovernmental Panel on Climate Change Expert Meeting on Mitigation, Sustainability and Climate Stabilization Scenarios*; Shukla, R.P., Skea, J., van Diemen, R., Calvin, K., Christophersen, Ø., Creutzig, F., Fuglestvedt, J., Huntley, E., Lecocq, F., Pathak, M., et al., Eds.; IPCC Working Group III Technical Support Unit, Imperial College London: London, UK, 2017.
4. Sobrino, J.A.; Oltra-Carrió, R.; Sòria, G.; Jiménez-Muñoz, J.C.; Franch, B.; Hidalgo, V.; Mattar, C.; Julien, Y.; Cuenca, J.; Paganini, M.; et al. Evaluation of the surface urban heat island effect in the city of Madrid by thermal remote sensing. *Int. J. Remote Sens.* **2012**, *34*, 3177–3192. [CrossRef]
5. Sánchez-Guevara Sánchez, C.; Núñez Peiró, M.; Neila González, F.J. Urban Heat Island and Vulnerable Population. The Case of Madrid. In *Sustainable Development and Renovation in Architecture, Urbanism and Engineering*; Springer International Publishing: Cham, Switzerland, 2017; pp. 3–13.
6. Taha, H. Characterization of Urban Heat and Exacerbation: Development of a Heat Island Index for California. *Climate* **2017**, *5*, 59. [CrossRef]
7. Oke, T.R. The energetic basis of the urban heat island. *Q. J. R. Meteorol. Soc.* **1982**, *108*, 1–24. [CrossRef]
8. Leal Filho, W.; Echevarria Icaza, L.; Neht, A.; Klavins, M.; Morgan, E.A. Coping with the impacts of urban heat islands. A literature based study on understanding urban heat vulnerability and the need for resilience in cities in a global climate change context. *J. Clean. Prod.* **2018**, *171*, 1140–1149. [CrossRef]
9. Khanian, M.; Serpoush, B.; Gheitarani, N. Balance between place attachment and migration based on subjective adaptive capacity in response to climate change: The case of Famenin County in Western Iran. *Clim. Dev.* **2019**, *11*, 69–82. [CrossRef]
10. Ling, T.-Y.; Chiang, Y.-C. Well-being, health and urban coherence-advancing vertical greening approach toward resilience: A design practice consideration. *J. Clean. Prod.* **2018**, *182*, 187–197. [CrossRef]
11. Singh, R.B.; Grover, A.; Zhan, J. Inter-Seasonal Variations of Surface Temperature in the Urbanized Environment of Delhi Using Landsat Thermal Data. *Energies* **2014**, *7*, 1811–1828. [CrossRef]
12. Aram, F.; Solgi, E.; Holden, G. The role of green spaces in increasing social interactions in neighborhoods with periodic markets. *Habitat Int.* **2019**, *84*, 24–32. [CrossRef]
13. Demuzere, M.; Orru, K.; Heidrich, O.; Olazabal, E.; Geneletti, D.; Orru, H.; Bhave, A.G.; Mittal, N.; Feliu, E.; Faehnle, M. Mitigating and adapting to climate change: Multi-functional and multi-scale assessment of green urban infrastructure. *J. Environ. Manag.* **2014**, *146*, 107–115. [CrossRef]
14. Zölch, T.; Maderspacher, J.; Wamsler, C.; Pauleit, S. Using green infrastructure for urban climate-proofing: An evaluation of heat mitigation measures at the micro-scale. *Urban For. Urban Green.* **2016**, *20*, 305–316. [CrossRef]

15. Park, J.; Kim, J.-H.; Lee, D.K.; Park, C.Y.; Jeong, S.G. The influence of small green space type and structure at the street level on urban heat island mitigation. *Urban For. Urban Green.* **2017**, *21*, 203–212. [CrossRef]

16. Buyadi, S.N.A.; Mohd, W.M.N.W.; Misni, A. Green Spaces Growth Impact on the Urban Microclimate. *Procedia Soc. Behav. Sci.* **2013**, *105*, 547–557. [CrossRef]

17. Lottrup, L.; Grahn, P.; Stigsdotter, U.K. Workplace greenery and perceived level of stress: Benefits of access to a green outdoor environment at the workplace. *Landsc. Urban Plan.* **2013**, *110*, 5–11. [CrossRef]

18. Imran, H.M.; Kala, J.; Ng, A.W.M.; Muthukumaran, S. Effectiveness of green and cool roofs in mitigating urban heat island effects during a heatwave event in the city of Melbourne in southeast Australia. *J. Clean. Prod.* **2018**, *197*, 393–405. [CrossRef]

19. Peng, L.L.; Jim, C.Y. Green-Roof Effects on Neighborhood Microclimate and Human Thermal Sensation. *Energies* **2013**, *6*, 598–618. [CrossRef]

20. Tan, C.L.; Wong, N.H.; Jusuf, S.K. Effects of vertical greenery on mean radiant temperature in the tropical urban environment. *Landsc. Urban Plan.* **2014**, *127*, 52–64. [CrossRef]

21. Li, J.; Zheng, B.; Shen, W.; Xiang, Y.; Chen, X.; Qi, Z. Cooling and Energy-Saving Performance of Different Green Wall Design: A Simulation Study of a Block. *Energies* **2019**, *12*, 2912. [CrossRef]

22. Kong, L.; Lau, K.K.-L.; Yuan, C.; Chen, Y.; Xu, Y.; Ren, C.; Ng, E. Regulation of outdoor thermal comfort by trees in Hong Kong. *Sustain. Cities Soc.* **2017**, *31*, 12–25. [CrossRef]

23. Lin, Y.-H.; Tsai, K.-T. Screening of Tree Species for Improving Outdoor Human Thermal Comfort in a Taiwanese City. *Sustainability* **2017**, *9*, 340. [CrossRef]

24. Wong, N.H.; Yu, C. Study of green areas and urban heat island in a tropical city. *Habitat Int.* **2005**, *29*, 547–558. [CrossRef]

25. Sugawara, H.; Shimizu, S.; Takahashi, H.; Hagiwara, S.; Narita, K.; Mikami, T.; Hirano, T. Thermal Influence of a Large Green Space on a Hot Urban Environment. *J. Environ. Qual.* **2016**, *45*, 125–133. [CrossRef]

26. Yang, C.; He, X.; Yu, L.; Yang, J.; Yan, F.; Bu, K.; Chang, L.; Zhang, S. The Cooling Effect of Urban Parks and Its Monthly Variations in a Snow Climate City. *Remote Sens.* **2017**, *9*, 1066. [CrossRef]

27. Qiu, G.Y.; Zou, Z.; Li, X.; Li, H.; Guo, Q.; Yan, C.; Tan, S. Experimental studies on the effects of green space and evapotranspiration on urban heat island in a subtropical megacity in China. *Habitat Int.* **2017**, *68*, 30–42. [CrossRef]

28. Lai, D.; Liu, W.; Gan, T.; Liu, K.; Chen, Q. A review of mitigating strategies to improve the thermal environment and thermal comfort in urban outdoor spaces. *Sci. Total Environ.* **2019**. [CrossRef]

29. Cao, X.; Onishi, A.; Chen, J.; Imura, H. Quantifying the cool island intensity of urban parks using ASTER and IKONOS data. *Landsc. Urban Plan.* **2010**, *96*, 224–231. [CrossRef]

30. Vidrih, B.; Medved, S. Multiparametric model of urban park cooling island. *Urban For. Urban Green.* **2013**, *12*, 220–229. [CrossRef]

31. Jamei, E.; Rajagopalan, P.; Seyedmahmoudian, M.; Jamei, Y. Review on the impact of urban geometry and pedestrian level greening on outdoor thermal comfort. *Renew. Sustain. Energy Rev.* **2016**, *54*, 1002–1017. [CrossRef]

32. Xu, L.; You, H.; Li, D.; Yu, K. Urban green spaces, their spatial pattern, and ecosystem service value: The case of Beijing. *Habitat Int.* **2016**, *56*, 84–95. [CrossRef]

33. Aram, F.; Higueras García, E.; Solgi, E.; Mansournia, S. Urban green space cooling effect in cities. *Heliyon* **2019**, *5*. [CrossRef] [PubMed]

34. Feyisa, G.L.; Dons, K.; Meilby, H. Efficiency of parks in mitigating urban heat island effect: An example from Addis Ababa. *Landsc. Urban Plan.* **2014**, *123*, 87–95. [CrossRef]

35. Lu, J.; Li, Q.; Zeng, L.; Chen, J.; Liu, G.; Li, Y.; Li, W.; Huang, K. A micro-climatic study on cooling effect of an urban park in a hot and humid climate. *Sustain. Cities Soc.* **2017**, *32*, 513–522. [CrossRef]

36. Anjos, M.; Lopes, A. Urban Heat Island and Park Cool Island Intensities in the Coastal City of Aracaju, North-Eastern Brazil. *Sustainability* **2017**, *9*, 1379. [CrossRef]

37. Yan, H.; Wu, F.; Dong, L. Influence of a large urban park on the local urban thermal environment. *Sci. Total Environ.* **2018**, *622–623*, 882–891. [CrossRef]

38. Aflaki, A.; Mirnezhad, M.; Ghaffarianhoseini, A.; Ghaffarianhoseini, A.; Omrany, H.; Wang, Z.-H.; Akbari, H. Urban heat island mitigation strategies: A state-of-the-art review on Kuala Lumpur, Singapore and Hong Kong. *Cities* **2017**, *62*, 131–145. [CrossRef]

39. Lin, P.; Gou, Z.; Lau, S.S.-Y.; Qin, H. The Impact of Urban Design Descriptors on Outdoor Thermal Environment: A Literature Review. *Energies* **2017**, *10*, 2151. [CrossRef]
40. ASHRAE Standard 55. *Thermal Environmental Conditions for Human Occupancy*; ASHRAE: Atlanta, GA, USA, 2010.
41. Coccolo, S.; Kämpf, J.; Scartezzini, J.-L.; Pearlmutter, D. Outdoor human comfort and thermal stress: A comprehensive review on models and standards. *Urban Clim.* **2016**, *18*, 33–57. [CrossRef]
42. Fanger, P.O. *Thermal Comfort: Analysis and Applications in Environmental Engineering*; Danish Technical Press: Copenhagen, Denmark, 1970.
43. Walls, W.; Parker, N.; Walliss, J. *Designing with thermal comfort Indices in Outdoor Sites. Paper presented at the Living and Learning: Research for a Better Built Environment: 49th International Conference of the Architectural Science Association 2015*; The Architectural Science Association and The University of Melbourne: Melbourne, Australia, 2015.
44. Bruse, M. Analysing human outdoor thermal comfort and open space usage with the multi-agent system BOTworld. In Proceedings of the 7th International Conference on Urban Climate, Yokohama, Japan, 29 June–3 July 2009.
45. Monteiro, L.A.; Alucci, M.P. Thermal Comfort Index for the Assessment of Outdoor spaces in Subtropical Climates. In Proceedings of the 7th International Conference on Urban Climate, Yokohama, Japan, 29 June–3 July 2009.
46. Potchter, O.; Cohen, P.; Lin, T.P.; Matzarakis, A. Outdoor human thermal perception in various climates: A comprehensive review of approaches, methods and quantification. *Sci. Total Environ.* **2018**, *631–632*, 390–406. [CrossRef]
47. Honjo, T. Thermal Comfort in Outdoor Environment. *Glob. Environ. Res.* **2009**, *13*, 43–47. [CrossRef]
48. Roshan, G.; Saleh Almomenin, H.; da Silveira Hirashima, S.Q.; Attia, S. Estimate of outdoor thermal comfort zones for different climatic regions of Iran. *Urban Clim.* **2019**, *27*, 8–23. [CrossRef]
49. Lin, T.-P.; Matzarakis, A.; Hwang, R.-L. Shading effect on long-term outdoor thermal comfort. *Build. Environ.* **2010**, *45*, 213–221. [CrossRef]
50. Santos Nouri, A.; Charalampopoulos, I.; Matzarakis, A. Beyond Singular Climatic Variables—Identifying the Dynamics of Wholesome Thermo-Physiological Factors for Existing/Future Human Thermal Comfort during Hot Dry Mediterranean Summers. *Int. J. Environ. Res. Public Health* **2018**, *15*, 2362. [CrossRef] [PubMed]
51. Chen, L.; Ng, E. Outdoor thermal comfort and outdoor activities: A review of research in the past decade. *Cities* **2012**, *29*, 118–125. [CrossRef]
52. Tseliou, A.; Tsiros, I.X.; Lykoudis, S.; Nikolopoulou, M. An evaluation of three biometeorological indices for human thermal comfort in urban outdoor areas under real climatic conditions. *Build. Environ.* **2010**, *45*, 1346–1352. [CrossRef]
53. Thorsson, S.; Lindberg, F.; Eliasson, I.; Holmer, B. Different methods for estimating the mean radiant temperature in an outdoor urban setting. *Int. J. Climatol.* **2007**, *27*, 1983–1993. [CrossRef]
54. Sun, S.; Xu, X.; Lao, Z.; Liu, W.; Li, Z.; Higueras García, E.; He, L.; Zhu, J. Evaluating the impact of urban green space and landscape design parameters on thermal comfort in hot summer by numerical simulation. *Build. Environ.* **2017**, *123*, 277–288. [CrossRef]
55. Chen, L.; Wen, Y.; Zhang, L.; Xiang, W.-N. Studies of thermal comfort and space use in an urban park square in cool and cold seasons in Shanghai. *Build. Environ.* **2015**, *94*, 644–653. [CrossRef]
56. Mahmoud, A.H.A. Analysis of the microclimatic and human comfort conditions in an urban park in hot and arid regions. *Build. Environ.* **2011**, *46*, 2641–2656. [CrossRef]
57. Bartesaghi Koc, C.; Osmond, P.; Peters, A. Evaluating the cooling effects of green infrastructure: A systematic review of methods, indicators and data sources. *Sol. Energy* **2018**, *166*, 486–508. [CrossRef]
58. Taleghani, M. Outdoor thermal comfort by different heat mitigation strategies—A review. *Renew. Sustain. Energy Rev.* **2018**, *81*, 2011–2018. [CrossRef]
59. Lenzholzer, S.; Klemm, W.; Vasilikou, C. Qualitative methods to explore thermo-spatial perception in outdoor urban spaces. *Urban Clim.* **2018**, *23*, 231–249. [CrossRef]
60. Nikolopoulou, M.; Baker, N.; Steemers, K. Thermal comfort in outdoor urban spaces: Understanding the human parameter. *Sol. Energy* **2001**, *70*, 227–235. [CrossRef]

61. Klemm, W.; Heusinkveld, B.G.; Lenzholzer, S.; Jacobs, M.H.; Van Hove, B. Psychological and physical impact of urban green spaces on outdoor thermal comfort during summertime in The Netherlands. *Build. Environ.* **2015**, *83*, 120–128. [CrossRef]

62. Knez, I.; Thorsson, S. Influences of culture and environmental attitude on thermal, emotional and perceptual evaluations of a public square. *Int. J. Biometeorol.* **2006**, *50*, 258–268. [CrossRef]

63. Nagashima, K.; Tokizawa, K.; Marui, S. Thermal comfort. *Handb. Clin. Neurol.* **2018**, *156*, 249–260. [CrossRef]

64. Thorsson, S.; Lindqvist, M.; Lindqvist, S. Thermal bioclimatic conditions and patterns of behaviour in an urban park in Goteborg, Sweden. *Int. J. Biometeorol.* **2004**, *48*, 149–156. [CrossRef]

65. Pezzoli, A.; Cristofori, E.; Gozzini, B.; Marchisio, M.; Padoan, J. Analysis of the thermal comfort in cycling athletes. *Procedia Eng.* **2012**, *34*, 433–438. [CrossRef]

66. Nasir, R.A.; Ahmad, S.S.; Ahmed, A.Z. Psychological Adaptation of Outdoor Thermal Comfort in Shaded Green Spaces in Malaysia. *Procedia Soc. Behav. Sci.* **2012**, *68*, 865–878. [CrossRef]

67. Lin, T.-P. Thermal perception, adaptation and attendance in a public square in hot and humid regions. *Build. Environ.* **2009**, *44*, 2017–2026. [CrossRef]

68. Nikolopoulou, M.; Lykoudis, S. Thermal comfort in outdoor urban spaces: Analysis across different European countries. *Build. Environ.* **2006**, *41*, 1455–1470. [CrossRef]

69. Ng, E.; Cheng, V. Urban human thermal comfort in hot and humid Hong Kong. *Energy Build.* **2012**, *55*, 51–65. [CrossRef]

70. Li, K.; Zhang, Y.; Zhao, L. Outdoor thermal comfort and activities in the urban residential community in a humid subtropical area of China. *Energy Build.* **2016**, *133*, 498–511. [CrossRef]

71. Yang, B.; Olofsson, T.; Nair, G.; Kabanshi, A. Outdoor thermal comfort under subarctic climate of north Sweden – A pilot study in Umeå. *Sustain. Cities Soc.* **2017**, *28*, 387–397. [CrossRef]

72. Klemm, W.; Heusinkveld, B.G.; Lenzholzer, S.; van Hove, B. Street greenery and its physical and psychological impact on thermal comfort. *Landsc. Urban Plan.* **2015**, *138*, 87–98. [CrossRef]

73. Lenzholzer, S. Microclimate perception analysis through cognitive mapping. In Proceedings of the PLEA 2008—25th Conference on Passive and Low Energy Architecture, Dublin, Ireland, 22–24 October 2008.

74. Xu, X.; Sun, S.; Liu, W.; García, E.H.; He, L.; Cai, Q.; Xu, S.; Wang, J.; Zhu, J. The cooling and energy saving effect of landscape design parameters of urban park in summer: A case of Beijing, China. *Energy Build.* **2017**, *149*, 91–100. [CrossRef]

75. Kottek, M.; Grieser, J.; Beck, C.; Rudolf, B.; Rubel, F. World Map of the Köppen-Geiger climate classification updated. *Meteorol. Z.* **2006**, *15*, 259–263. [CrossRef]

76. AEMET. Agencia Estatal de Meteorología. 2018. Available online: http://www.aemet.es/es/portada (accessed on 11 October 2018).

77. Núñez Peiró, M.; Sánchez-Guevara Sánchez, C.; Neila González, F.J. Update of the Urban Heat Island of Madrid and Its Influence on the Building's Energy Simulation. In *Sustainable Development and Renovation in Architecture, Urbanism and Engineering*; Springer International Publishing: Cham, Switzerland, 2017; pp. 339–350.

78. Román, E.; Gómez, G.; de Luxán, M. Urban Heat Island of Madrid and Its Influence over Urban Thermal Comfort. In *Sustainable Development and Renovation in Architecture, Urbanism and Engineering*; Springer International Publishing: Cham, Switzerland, 2017; pp. 415–425.

79. ABIO. Arquitectura Bioclimática en un Entorno Sostenible (Research Group Was Officially Recognized by the Universidad Politécnica de Madrid). 2018. Available online: http://abio-upm.org/ (accessed on 17 May 2018).

80. Morakinyo, T.E.; Kong, L.; Lau, K.K.-L.; Yuan, C.; Ng, E. A study on the impact of shadow-cast and tree species on in-canyon and neighborhood's thermal comfort. *Build. Environ.* **2017**, *115*, 1–17. [CrossRef]

81. Fröhlich, D.; Gangwisch, M.; Matzarakis, A. Effect of radiation and wind on thermal comfort in urban environments—Application of the RayMan and SkyHelios model. *Urban Clim.* **2019**, *27*, 1–7. [CrossRef]

82. Matzarakis, A.; Rutz, F.; Mayer, H. Modelling radiation fluxes in simple and complex environments—Application of the RayMan model. *Int. J. Biometeorol.* **2007**, *51*, 323–334. [CrossRef]

83. Auliciems, A.; Szokolay, S.V. *Thermal Comfort*, 2nd ed.; Szokolay, S.V., Ed.; PLEA in Association with Dept. of Architecture, University of Queensland: Brisbane, QLD, Australia, 2007.

84. ASHRAE Standard 55. *Thermal Environmental Conditions for Human Occupancy*; ASHRAE: Atlanta, GA, USA, 2013.

85. Matzarakis, A.; Rutz, F.; Mayer, H. Modelling radiation fluxes in simple and complex environments: Basics of the RayMan model. *Int. J. Biometeorol.* **2010**, *54*, 131–139. [CrossRef] [PubMed]

86. Maria Raquel, C.d.S.; Montalto, F.A.; Palmer, M.I. Potential climate change impacts on green infrastructure vegetation. *Urban For. Urban Green.* **2016**, *20*, 128–139. [CrossRef]

87. Yen, Y.; Wang, Z.; Shi, Y.; Xu, F.; Soeung, B.; Sohail, M.T.; Rubakula, G.; Juma, S.A. The predictors of the behavioral intention to the use of urban green spaces: The perspectives of young residents in Phnom Penh, Cambodia. *Habitat Int.* **2017**, *64*, 98–108. [CrossRef]

88. Aram, F.; Solgi, E.; Higueras García, E.; Mohammadzadeh, S.; Mosavi, A.; Shamshirband, S. Design and Validation of a Computational Program for Analysing Mental Maps: Aram Mental Map Analyzer. *Sustainability* **2019**, *11*, 3790. [CrossRef]

89. Barnett, V.; Neter, J.; Wasserman, W. Applied Linear Statistical Models. *J. R. Stat. Soc. Ser. A (Gen.)* **2006**, *138*, 258. [CrossRef]

90. Lawrence, I.; Lin, K. A concordance correlation coefficient to evaluate reproducibility. *Biometrics* **1989**, 255–268.

91. Akoglu, H. User's guide to correlation coefficients. Turkish Journal of Emergency Medicine. *Emerg. Med. Assoc. Turk.* **2018**. [CrossRef]

92. Pereira, P.F.; Ramos, N.M.M. Occupant behaviour motivations in the residential context – An investigation of variation patterns and seasonality effect. *Build. Environ.* **2019**, *148*, 535–546. [CrossRef]

93. Santos Nouri, A.; Fröhlich, D.; Matos Silva, M.; Matzarakis, A. The Impact of Tipuana tipu Species on Local Human Thermal Comfort Thresholds in Different Urban Canyon Cases in Mediterranean Climates: Lisbon, Portugal. *Atmosphere* **2018**, *9*, 12. [CrossRef]

94. Laureti, F.; Martinelli, L.; Battisti, A. Assessment and Mitigation Strategies to Counteract Overheating in Urban Historical Areas in Rome. *Climate* **2018**, *6*, 18. [CrossRef]

95. Skoulika, F.; Santamouris, M.; Kolokotsa, D.; Boemi, N. On the thermal characteristics and the mitigation potential of a medium size urban park in Athens, Greece. *Landsc. Urban Plan.* **2014**, *123*, 73–86. [CrossRef]

96. Oliveira, S.; Andrade, H.; Vaz, T. The cooling effect of green spaces as a contribution to the mitigation of urban heat: A case study in Lisbon. *Build. Environ.* **2011**, *46*, 2186–2194. [CrossRef]

97. Cohen, P.; Potchter, O.; Matzarakis, A. Daily and seasonal climatic conditions of green urban open spaces in the Mediterranean climate and their impact on human comfort. *Build. Environ.* **2012**, *51*, 285–295. [CrossRef]

Article

Mirroring Solar Radiation Emitting Heat Toward the Universe: Design, Production, and Preliminary Testing of a Metamaterial Based Daytime Passive Radiative Cooler

Anna Castaldo [1,*], Giuseppe Vitiello [1], Emilia Gambale [1], Michela Lanchi [2], Manuela Ferrara [3] and Michele Zinzi [4]

[1] ENEA-TERIN-STSN-SCIS Portici, 80055 Naples, Italy; giuseppe.vitiello@enea.it (G.V.); emilia.gambale@enea.it (E.G.)
[2] ENEA-TERIN-STSN-SCIS Casaccia, 00123 Rome, Italy; michela.lanchi@enea.it
[3] ENEA-TERIN-FSN-DIN Portici, 80055 Naples, Italy; manuela.ferrara@enea.it
[4] ENEA-TERIN-SEN Casaccia, 00123 Rome, Italy; michele.zinzi@enea.it
* Correspondence: anna.castaldo@enea.it; Tel.: +39-(0)81-772-3358

Received: 16 June 2020; Accepted: 10 August 2020; Published: 13 August 2020

Abstract: A radiative cooling device, based on a metamaterial able to mirror solar radiation and emit heat toward the universe by the transparency window of the atmosphere (8–13 µm), reaching and maintaining temperatures below ambient air, without any electricity input (passive), could have a significant impact on energy consumption of buildings and positive effects on the global warming prevention. A similar device is expected to properly work if exposed to the nocturnal sky, but during the daytime, its efficacy could be affected by its own heating under direct sunlight. In scientific literature, there are only few evidences of lab scale devices, acting as passive radiative cooling at daytime, and remaining few degrees below ambient air. This work describes the proof of concept of a daytime passive radiative cooler, entirely developed in ENEA labs, capable to reach well 12 °C under ambient temperature. In particular, the prototypal device is an acrylic box case, filled with noble gas, whose top face is a metamaterial deposited on a metal substrate covered with a transparent polymeric film. The metamaterial here tested, obtained by means of a semi-empirical approach, is a spectrally selective coating based on low cost materials, deposited as thin films by sputtering on the metallic substrate, that emits selectively in the 8–13 µm region, reflecting elsewhere UV_VIS_NIR_IR electromagnetic radiation. The prototype during the daytime sky could reach temperatures well beyond ambient temperature. However, the proof of concept experiment performed in a bright clear June day has evidenced some limitations. A critical analysis of the obtained experimental results has done, in order to individuate design revisions for the device and to identify future metamaterial improvements.

Keywords: cool roof; passive radiative cooling; metamaterials; prototype

---

1. Introduction

The average ambient air temperature has raised by about 0.12 °C, per decade, since 1951 because of climate change [1]. This trend has been emphasized at a local scale by the Urban Heat Island (UHI) effect, defined as the increase of urban temperature compared to that of the countryside surrounding the city, caused by the synergic action of urban sprawl and global warming [2–4].

Aligned with the ambient temperature increase, a relevant rise of the electricity consumption can be expected in the future [5]. Wenz et al. [6] demonstrated that electricity uses in Southern European countries exceeded by 7% that of Northern European ones, due to the occurrence of higher temperature levels often concentrated in short periods, attributable to building space cooling [7]. Implications also regard thermal comfort and quality of life, especially for low-income people [8].

The role of proper building materials in the mitigation strategy to fight against overheating is widely documented in the literature [9–11]. The so-called cool materials, which are characterized by high solar reflectance and high thermal emissivity, can limit the temperature rise of urban fabrics. Several technologies emerged in the past years to enhance the building energy performance and architectural integration: spectrally selective [12], fluorescent [13], dynamic [14] materials. Moreover, the production and characterization of super cool materials is documented in [15]; these materials remain below the ambient temperature under solar irradiation, thanks to extremely high emissivity and solar reflectance (both above 0.96/0.97).

In this perspective, potentialities of daytime radiative coolers are very high and, thus, they are gaining attention as ways to cool down buildings and cities. Benefits of nighttime radiative cooling have been documented in recent decades [16,17], while daytime radiative cooling is an emerging technology, which has to face the challenging task of keeping the material surface well below the ambient temperature under solar irradiation [18]. Few materials can fulfill this objective thanks to very high reflectance in all the solar and in infrared range with exception of the atmosphere transparency window (8–13 μm), which also corresponds to the peak thermal emission of bodies at ambient temperature [19]. These materials must have very high emissivity only in the latter range and are, thus, able to cool down radiating the heat in the outer space [20].

Despite significant progress in the field, daytime cooling technologies are still under study and development [21]. Recent advances in the electromagnetic field, related to nanotechnologies, have demonstrated to be promising for this specific application [19]. Essentially a photonic approach can modulate the emission and adsorption of electromagnetic radiations in specific ranges, to create one-dimensional (1D), 2D, and 3D metamaterials [22]. A first application of a photonic approach to obtain a device with a metamaterial able to reach 5 °C under ambient temperature during daytime has been reported by Raman and colleagues of the Stanford team, in 2014 [23]. Their proposed metamaterial is an alternating dielectric layers stack, obtained based on hafnium and silicon oxides deposited by evaporation. Other solutions based on metamaterials or porous alumina were developed more recently [24–26].

In consideration of the possible closure of the atmosphere transmittance window in presence of high humidity (water vapor blocks infrared radiation), many studies have been done on the evaluation of the influence of emission selectivity [27]. For example, in determined environmental conditions, reflecting only solar range and emitting elsewhere could result simplest and cheaper [28]. Moreover, polymeric formulations containing silica nano-spheres to be coupled with metallic reflective layers [29] have been proposed and compared with more sophisticated photonic metamaterials.

Starting from a multi-year experience in the development of spectrally selective coatings for receiver tubes in solar thermodynamic field [30], the authors have recently started an experimental activity focused on the development of simple, low cost, scalable coatings for daytime radiative cooling to be integrated in buildings fabrics or cooling system components. The aim of this work is to contribute to the advancement of knowledge and current state of art and research on this field, both in terms of new materials development and prototypal device fabrication, highlighting the following aspects:

- design and fabrication of new metamaterials on different substrates, pursuing a photonic approach similar to that reported from Raman and colleagues, in terms of reflectance higher than 95% in all solar spectral range, and selective emission only in 8–13 micron, but with cheaper materials and utilizing a different deposition process in spite of evaporation, as the low cost and scalable sputtering deposition;
- realization of a daytime radiative cooling prototypal device, for individuating possible applications and testing cool metamaterial and/or other similarly produced materials, which, taking into right account different contribution to heat transfer (conduction, convection, radiation, etc.),

can maximize the optical performance of selective emission in 8–13 micron and reflective behavior in all wavelength range, working as heat exchanger if deposited onto metallic substrates.

Proof of concept (on technological scale) of the use of metamaterials here developed to reach a temperature well below ambient air during daytime.

Objective of the present study is the design and preparation of a daytime radiative cooler, characterized by an extremely high reflectance in all wavelengths except in the atmospheric transparency window (8–13 micron); thus, able to remain well below ambient temperature under solar irradiation. The use of cheaper materials and the sputtering deposition process, instead of evaporation, may lead to lower costs compared to similar developed solution

In particular, the proof of concept of the daytime passive radiative cooler device has been performed either protecting it by a transparent case, either directly exposing it to the outdoor environment, aiming at assessing the surface temperature drop potentialities in the Mediterranean climate and the options to implement low cost solutions.

In the following sections, the design and manufacturing of the prototype, the design and manufacturing of metamaterials, the characterization of chemical-physical and optical properties of the metamaterial, the testing on field of the prototype will be described.

2. Materials and Methods

The present daytime cooling prototypal device consists of a metamaterial deposited on a metallic substrate, directly exposed to the outdoor environment, or protected by a dedicated transparent case filled with a noble gas. Possible applications are: i) coupling with a cooling system, whose pipes run below the radiator (active/hybrid mode); ii) directly integrated in the building envelope as radiative cooler (passive mode), iii) simply utilization to test new selective engineered materials.

The present study is limited to the surface temperature drop with respect to the ambient temperature, as the unique performance indicator. Next developments will include the cooling power as additional indicator, to compare the results versus the already developed research and technology in this regard.

2.1. Design and Manufacturing of the Prototype

The case prototype has been specifically designed for this activity, and entirely realized in ENEA labs. It is a rectangular box 500 × 400 × 115 mm, made of acrylic sheets nominally 10 mm thick and filled with pure Krypton, to minimize heat convection. The total weight of the box is about 6 Kg.

The prototype has a 300 × 200 mm exposed area and it consists of the following parts, as indicated in the left scheme in Figure 1:

- transparent cover, to insulate the metamaterial from wind and water condensation;
- base layer, made of the acrylic sheet with a layer of insulating polystyrene foam placed on, and where the passive radiative cooler plate is collocated. This suppresses the conductive heat gain from the back surface and side-walls;
- connector for the installed thermocouple, used to measure the surface temperature during the test;
- screws to fix the transparent cover onto the base;
- valve to inflate gas into the case;
- valve to exit gas from the case.

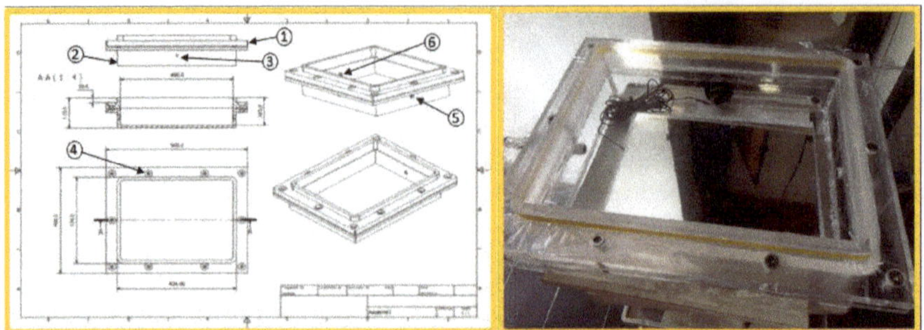

Figure 1. Cooler prototype scheme and photo.

2.2. Design and Manufacturing of the Cool Metamaterial

The radiative cooler consists of a substrate of aluminum of dimensions $300 \times 200 \times 1$ mm coated with an engineered material, named in following sections cool metamaterial (CM) fabricated in ENEA labs by means of sputtering deposition technique.

In particular, the present CM has been obtained by means of a photonic approach (semi-empirical simulation method), already described in our previous works [31]. In more detail, a sequence of layers with specific thicknesses has been selected, based on our consolidated knowledge in the field of dielectric materials. Through the application of a commercial optical properties simulation software, Essential Macleod (a software for the design, analysis, simulation and troubleshooting of thin optical coatings), a first sequence of layers can be individuated starting from the data-base of the software and imposing desired transmittance or reflectance values to the entire structure. A second step consists in depositing each single layer on a glass substrate and in measuring the optical properties through spectrophotometric and ellipsometric experimental analysis. The last step consists in including the obtained experimental data in the software library, iterating simulation of the entire structure to obtain the theoretical optimal method (thickness of layers and their right sequence) to realize the CM with desired optical properties by means of sputtering. However, sequential deposition of different materials can create compatibilities issues that requires the addition of new layers or the modification of initially selected materials, iterating the previous steps until the optimization of the fabrication method. Finally, chemical-physical characterization of the experimental metamaterial can confirm the achievement of the desired optical properties or give suggestions on further iteration of the previous steps.

The CM selected for this work is part of a library of engineered metamaterials that can be optimized for different substrates. In particular, it consists of different layers of selected materials sputtered in the following conditions:

- silver (Ag) film, grown in DC sputtering mode, at Ar pressure of 0.1 Pa and with 1.65 W/cm^2 power density applied to the cathode;
- aluminum nitride (AlN) layers, grown at Ar+N$_2$ pressure of 1 Pa and with power density of 7.75 W/cm^2 applied to Al cathode;
- aluminum oxide (Al$_2$O$_3$) layer, deposited at Ar+O$_2$ pressure of 1 Pa and with power density of 7.75 W/cm^2 supplied to Al cathode;
- silicon nitride (SiN$_x$) films grown in a gas mixture of Ar+N$_2$ at different concentration and total pressure ranging from 0.7 to 1 Pa. The Si target has been pretreated using an Ar+H$_2$ gas mixture to reduce the contamination on the Si target surface. As an example for SiN$_x$_01 and SiN$_x$_02 these experimental conditions have been applied:
- SiN$_x$_01 medium frequency 200W, 200sccm Ar+, 30 sccm of N$_2$ 0.7 Pa;
- SiN$_x$_02 medium frequency 1700 W, 200sccm Ar+, 30 sccm of N$_2$ 1Pa.

UV-VIS-NIR analysis has been performed using a double beam Perkin-Elmer model. Lambda 900 instrument, equipped with a 15 cm integrating sphere to measure global spectral reflectance and transmittance.

Ellipsometric analysis has been executed using a J.A. Woollam variable-angle spectroscopic instrument (Model VB-400) equipped with WVASE32 software.

Vibrational analysis of materials has been done using two Fourier Transform Infrared (FTIR), a Perkin Elmer for transmittance measurements, equipped with a Deuterated Triglycine Sulfate (DTGS) detector operating in the 400–10,000 cm^{-1} range and a Bruker Equinox with integrating sphere and MCT detector operating in the 760–8000 cm^{-1} range for diffuse reflectance measurements. The resolution has been comprised in the range 0.5 to 4 cm^{-1} for transmittance and 8 cm^{-1} for reflectance.

Substrates used for characterization purposes are Si (111), optical quality glass and stainless steel. As for the prototype fabrication, the selected substrate is aluminum, for a potential better integration in heat exchanger applications.

2.3. Field Testing

Experiments on field have been conducted in a sunny June 2019 day, comparing different temperatures read from thermocouples positioned as in Scheme 1 and recording data with a ThermoCamera. The thermocouples used to monitor T are Watlow shim style mod. P/N 75XKSGA180B and model. P/N 70XKSGD180B with a stainless steel shim, which can also be placed between components for surface temperature measurement. The ThermoCamera is Fluke Ti27 IR FUSION TECHNOLOGY.

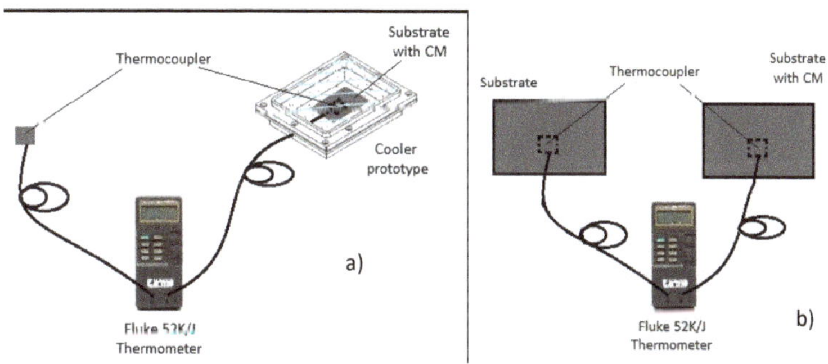

Scheme 1. Temperature test experiments on field.

The images are recorded, analyzed with the SmartView software, and compared with the data from the thermocouple, as reported in Figure 2.

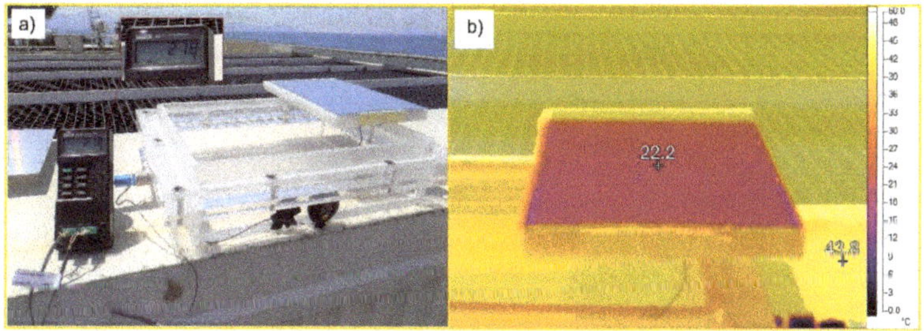

Figure 2. Thermocouple (**a**) and thermocamera (**b**) reading temperature comparison.

This validation test can be useful to assess, in the temperature range under exam, an error, in the temperature measurement, of the order of +/− 0.5° C.

The environmental average conditions of the day of the experiment are the following: wind speed = 3.11 +/− 0.08 m/s; wind direction 214 +/− 1° 0° = North; T = 26.41 +/− 0.11 °C.

3. Results

The proof of concept of a passive daytime radiative cooling device, potentially integrable with a chiller or a fan coil, has required the following steps: 1) fabrication of the prototype case; 2) fabrication and characterization of the emissive metamaterial, CM; 3) selection of cover polymeric material, which preserves optical properties of CM; 4) field testing of the device operation as passive radiative cooler.

The design and realization of the prototype case is described in the previous paragraph and can be considered as a first important practical result.

The second step is the development of an engineered cool metamaterial, CM, able to satisfy some stringent optical requirements (e.g., selective emissivity) as reported in Figure 3, and the deposition of it on an aluminum plate acting as heat exchanger, by means of reproducible, scalable and low cost processes.

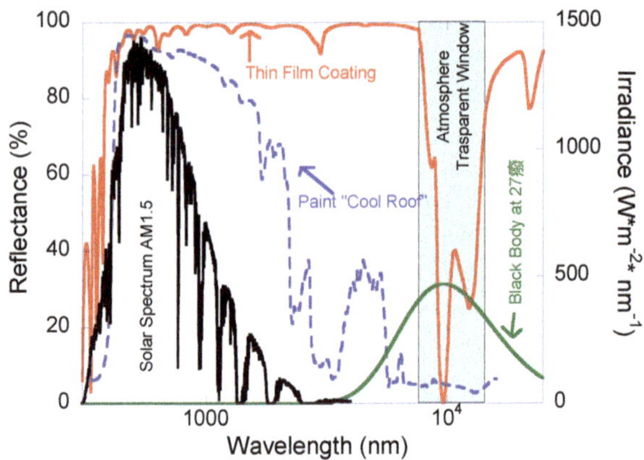

Figure 3. Reflectance spectra in the wavelength range 250 to 16,000 nm of simulated multilayered cool metamaterial (CM) (red line) in comparison to a simple cool roof paint (blue line), reported with Solar Spectrum AM1.5 irradiance (dark line) and black body irradiance at 27 °C (green line).

CM has been obtained by means of a photonic approach (semi-empirical simulation method), already described in ENEA's previous works, and detailed in the above section. In particular, a sequence of layers with specific thicknesses has been selected, based on a consolidated knowledge in the field of dielectric materials, and the resulting metamaterial properties have been simulated in solar range using a commercial optical properties simulation software (McLeod), to grant an high reflectance in such spectral region.

The resulting CM consists of different layers of silver, aluminum, and silicon oxides and nitrides (SiN_x, SiO_2, Al_2O_3, AlN), with variable thicknesses. Two main requirements have been considered for the material definition: 1) strong emissivity in 8–13 micron, 2) different adaptive template layers, tailored for all different substrates where the metamaterial has to be deposited, both for development/characterization purpose and for prototype plate fabrication. In Table 1, the main structure obtained is reported.

Table 1. Cool metamaterial (CM) structure composition.

Layer Composition	Thickness [nm]
SiO_2	230
SiN_x	150
SiO_2	230
Al_2O_3	150
SiN_x	100
Al_2O_3	360
AlN	100
Ag	200
AlN	10
Substrate	>1500

The silicon based layers mainly contributed to the IR emissivity in the 8–13 μm wavelengths range while the thinner alumina bottom layers deposited on Ag coated metallic substrate, work in maximizing (>95%) solar reflection.

Once defined, the materials sequence and the thicknesses of each layer, a relevant effort has been spent to fabricate CM. In particular, the work consisted in a full process manufacturing optimization, firstly requiring the sputtering deposition and characterization of each single layer, and then the full coating sequential deposition.

For a comprehensive chemical-physical characterization, the CM has been deposited on different substrates (stainless steel, silicon, glass), and on aluminum foil of 10 × 30 cm. To cover metallic substrates with the reflective silver layer, which has the highest reflectance in solar range (400–2500 nm), it is preferable to use an adaptive layer because metallic substrates are (usually) covered with native oxides, whereas silver growth on oxides is difficult [32]. To this purpose the consolidated technology of sandwiching silver between aluminum nitride layers [33] has been adopted; the first AlN layer has the function of barrier toward diffusion of stainless steel elements and silver template, while the second aluminum nitride layer is a barrier between silver and oxygen contained in alumina successive layer. Another kind of adaptive layer can be Tungsten alpha or other materials that can work as template for a 2D of the noble metal at low thickness.

Alumina last layer was deposited by means of reactive sputtering, from an aluminum target and an oxygen reactive gas in target saturation region.

The role of silicon nitride is crucial for granting high selective emissivity. Within the present work, SiN_x was deposited by means of reactive sputtering in N_2 and Argon atmosphere, working in transition region of the silicon target. It is typically very difficult to sputter crystalline Si_3N_4 (bulk material). In order to obtain the simulated metamaterial formulation, two silicon nitrides, with Si/N ratio < 1.3, were produced and characterized. In Figure 4, refractive index of the two samples SiN_x01 and SiN_x02 are reported, to show how it is possible to act on optical properties of thin layers varying sputtering process parameters.

In Figure 5 is shown a comparison between infrared reflectance of CM simulated and CM deposited on aluminum plate.

In Figure 6, a comparison between CM FTIR reflectance of CM, as deposited on different substrates by means of the same sputtering process is reported. It is remarkable the high reflectance in a large part of infrared range (>95% from 2000 to 6000 cm^{-1}) and the strong absorption (and, hence, emission) in 670–1300 cm^{-1}. Only few differences between metamaterials, as grown on different substrates in infrared reflectance spectra, are detectable, because reflective behavior of such metamaterial in this range is heavily influenced by the 200 nm silver layer. The absorption band (670–1300 cm^{-1}) is identical in all different samples. A difference can be detected in the hydrogen bonds absorption band around 3300 cm^{-1}, lied to –OH groups and around 3600 cm^{-1} lied to –NH groups and influenced by water content.

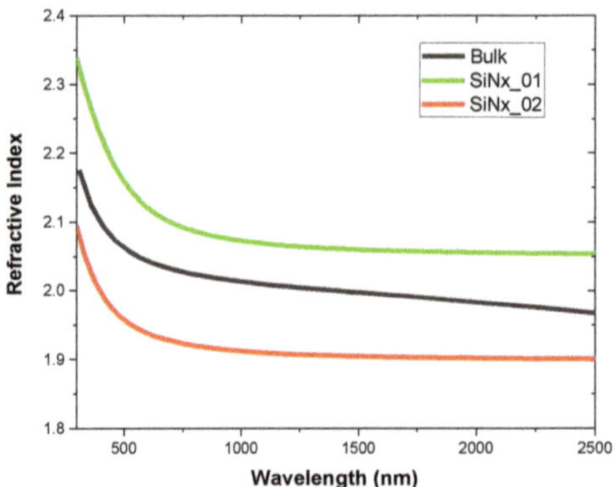

Figure 4. Refractive index of two silicon nitride thin layers compared to bulk one.

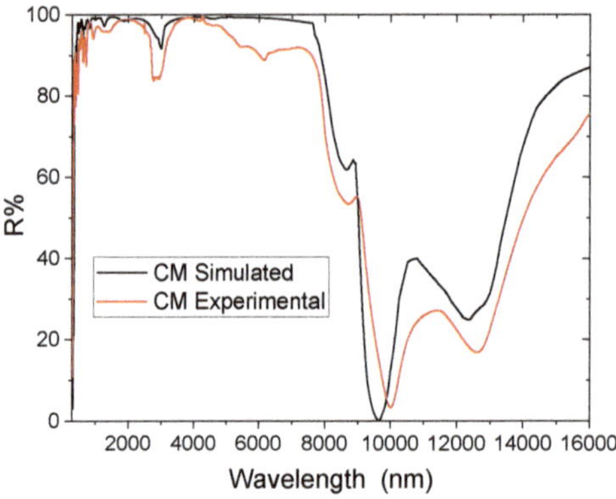

Figure 5. Infrared reflectance of CM simulated and experimental.

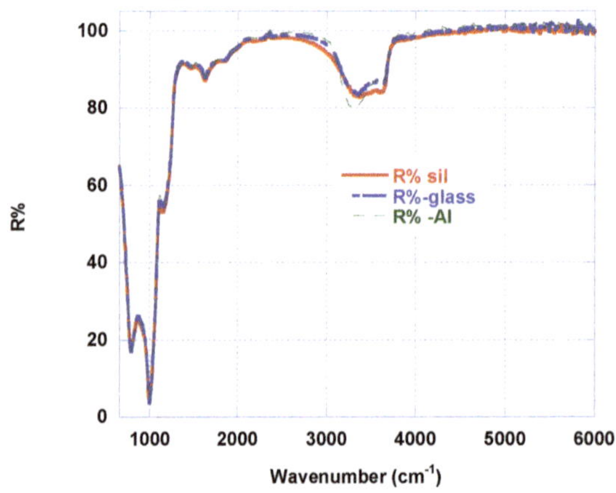

Figure 6. Infrared reflectance of CM deposited on different substrates: silicon (red line), glass (blue line), aluminum (green line).

Apart from CM, a cover protecting the system from natural convection can also play an important role to optimize the performance of the radiative cooler. As above mentioned, for daytime application, it is fundamental to block convective heat gain, so that the cooling prototype's net radiated power is higher than the absorbed solar power and the device can yield a temperature below the ambient one. On the contrary, if the absorbed solar power is higher than the net radiated power, the device temperature becomes higher than the ambient air temperature. In this circumstance, the use of convective covers is undesirable, as natural convection can facilitate the release of heat from the cooling device. In the absence of direct solar radiation, convection covers can greatly improve the device performance in terms of radiative cooling.

To cover the box, protect the panel from air and wind, and suppress convective heat gain from the ambient, it is necessary to individuate a material with a high optical transmittance in 8–13 micron range, since differently this could reduce CM emissivity.

Moreover, the covering material has to be flexible and low cost. An organic material typically absorbs in such a region, but in literature it has been demonstrated that a thin film of pure polyethylene has a transmittance higher than 80% in the IR region 8–13 μm (769–1250 cm^{-1}), as reported in Figure 7a (book transmittance spectrum of polyethylene card). Different commercial samples of PE (of various origins and thicknesses) have been analyzed with other polymeric transparent materials (e.g., Fluon of different type, PE-PET of different thickness, PC, etc.). As an example, in Figure 7b, the FTIR Transmittance percentage (T%) vs. wavenumber is reported for different selected samples (Fluon 250 (red line), PE film 6 micron thick (blue line), PET-PE 10 micron thick (green line), PE based freezer bag (dark line). Focusing on the interest region (Figure 7c), a 6-micron thick PE film cab be considered appropriate for the purposes of the present work. It is worth noticing that, in some commercial PE based polymers, a white pigment is present that interferes with optical behavior. In other terms, it is always advisable to verify, through FTIR analysis that the covering film is transparent in the specific interest region. Differently, the work to optimize metamaterial optical properties could be useless, since the effective front of the device is the polymer.

As shown, the selected PE film (6 micron thick) has the highest IR transmittance (>80%) in the range 8–13 micron (769–1250 cm^{-1}). As already mentioned, inside the rectangular box, the gap between stainless and polymeric cover has been filled with the noble gas Krypton, which has the lowest heat transfer coefficient, for reducing convection heat transfer.

Verification of the Functionality of the Prototype

The first experimental test on the prototype was conducted in June 2019, on a rooftop of the ENEA Portici Lab. In a preliminary phase of the test, the prototype was positioned on a rooftop, with thermocouples measuring metal radiator temperature and air temperature (as described in Scheme 1a): no temperature decrease of CM was observed, rather a little overheating. Subsequently, the polymeric cover was removed from the box, to avoid greenhouse effect, and the experiment was repeated on the metallic plate exposed directly to air, positioned on the casing but without cover. The plate temperature, measured on the underside of the heat exchanger, after 2 min went well below the air temperature, approximately of 12 °C, even under direct solar irradiance staying unchanged for the duration of the test (1 h), as shown in Figure 8.

It can be hypothesized that the heat gain of the device, in the preliminary part of the test, was due to the direct sun irradiation of the box lateral sides. Sunshades should have been mounted around the periphery of the enclosure, to prevent the sun from heating up the box walls, causing greenhouse effect. Most probably, sunshades are not required for night operation of the device. The simple removal of the coverage was effective in reducing the device temperature. Since CM was directly exposed to air, the system did not perform at the best of its potentiality: indeed the presence of inert gas and PE film is expected to significantly contribute to reduce heat gain from ambient convection.

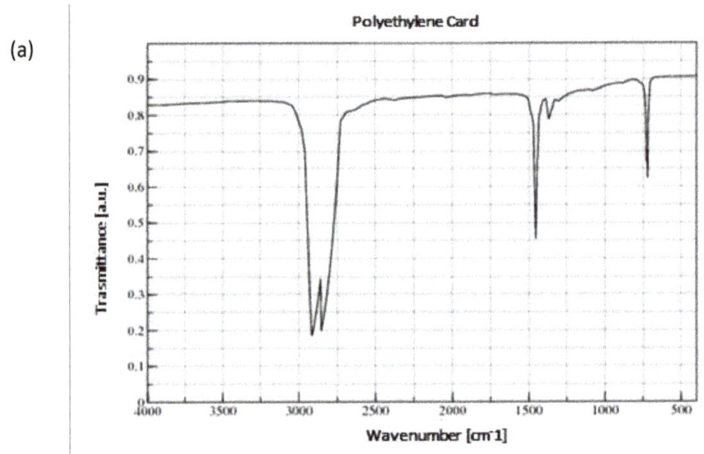

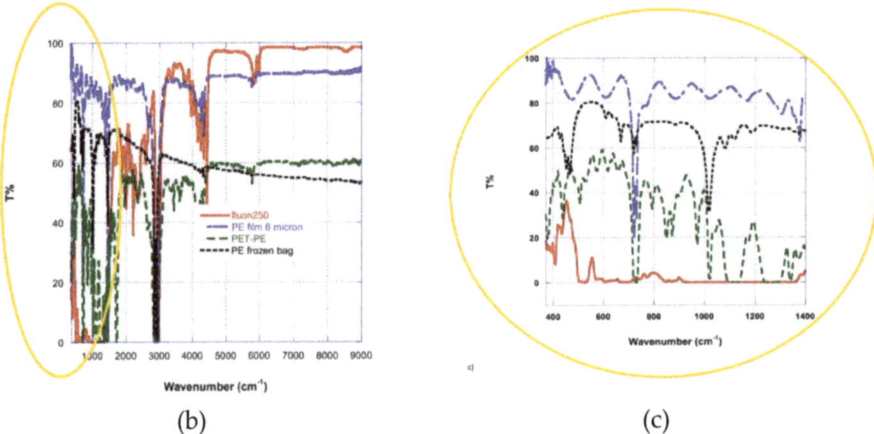

Figure 7. FTIR transmittance spectra of different polymeric thin films for selecting cover material. (**a**) Book spectrum of polyethylene card; (**b**) full range spectra of Fluon 250 (red line), PE film 6 micron thick (blue line), PET-PE (green line), PE based freezer bag (dark line). (**c**) Insert containing the interest region (769–1250 cm^{-1}), where the most transparent polymer is the blue-one line.

Anyway, having extracted CM from the box (and consequently having only measured the CM thermal behavior and not the entire prototype performance), another shorter test was executed to confirm the metamaterial ability to reach temperatures lower than the ambient one. In particular, the temperature difference (ΔT) occurring between two metallic identical plates exposed to clear sky conditions, one of which covered with metamaterial, was measured positioning two thermocouples as represented in Scheme 1b. In Figure 9, the maximum temperature difference measured (a) and the temperature course (b) are reported. As shown in Figure 9, the CM was able to reach a temperature 12 °C lower than its substrate.

The CM exposed to the air after the test started to partly peel off from the metallic slab substrate edges. The peeling process that initially started in the edges region (Figure 10b), after 1 month of air exposure (Figure 10a), appeared uniform on the entire slab. Pink regions, entirely delaminated from substrate, are visible in Figure 10a. Such behavior is not detectable in samples deposited on different substrates (produced for analytical purposes).

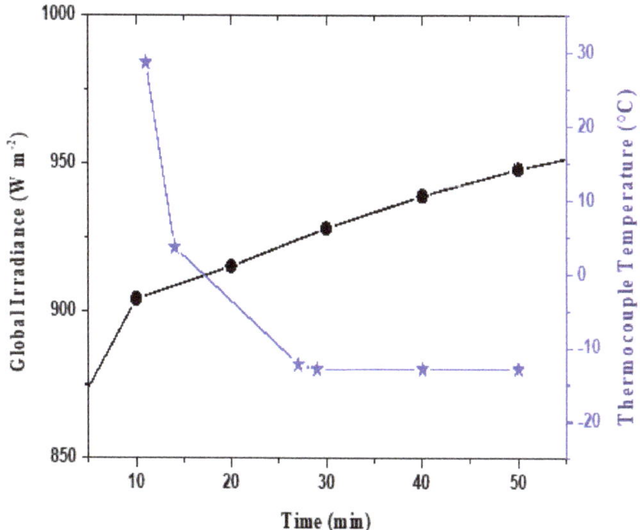

Figure 8. Experimental characterization of the radiative cooler prototype on the roof: first experimental test.

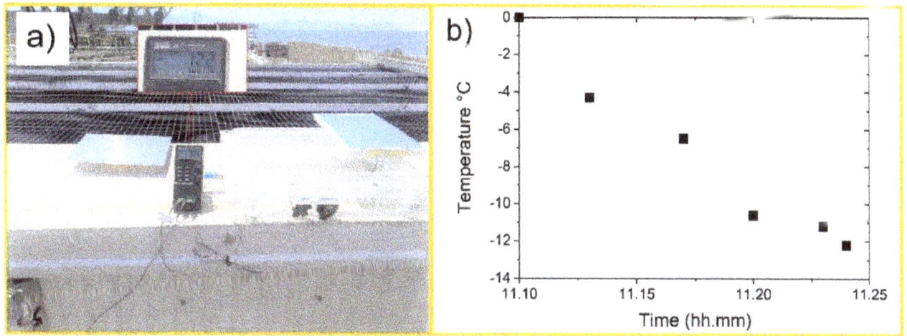

Figure 9. Comparative temperature test between CM and identical plate without coating exposed to clear sky conditions: (**a**) photo, (**b**) data.

The FTIR reflectance spectrum of the metallic sample after 1 month of air exposure was measured and it is reported in Figure 11, compared to as grown one. A degradation phenomenon that reduces emission performances in the transparent atmosphere window is evident. In particular, a decrease of TO and LO Si-N and Si-O absorption modes is present.

a)

b)

Figure 10. (**a**) Photos of mirror coatings before and after test. (**b**) ThermoCamera image during test.

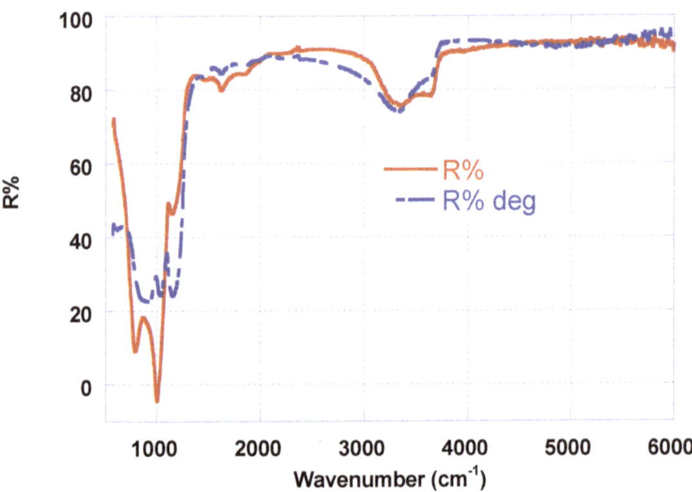

Figure 11. FTIR Reflectance before the field test and after 1 month.

4. Discussion

Aims of this work is the proof of concept of a passive daytime radiative cooling device, to be potentially integrated with a chiller or a fan coil (and in such case semi-passive) to maintain cool an indoor environment reducing electricity consumption. This electricity saving can be achieved if the cooler located on the roof is able to grant, under direct sunlight, a temperature well below ambient temperature.

In a classical macroscopic physics approach, a radiative cooler exposed to a daylight sky is subject to both global solar irradiance and atmosphere thermal radiation. The net cooling power emitted by the radiator depends on the power balance between radiative, convective, and conductive transport phenomena. In particular, the device radiative power has to increase, while the conductive and convective parts have to lower. The radiative device power can be maximized with the use of emissive materials (that strongly emit in the window transparency window of 8–13 micron range). The amount of the incidental atmospheric radiation ($P_{sunAM1.5}$), which is absorbed by the radiator, can be diminished by increasing reflectance in solar range (250–2500 nm). The power absorbed by the surrounding atmosphere (P_{atm}) and the power gained due to convective and conductive phenomena ($P_{conv+cond}$), which could heat the device can be lowered by means of a proper insulation from ambient.

When the radiated output power exceeds the net absorbed power ($P_{sunAM1.5} + P_{atm} + P_{conv+cond}$), the radiator temperature lowers.

Thus, to achieve a state where T is significantly below T_a, an engineered material with peculiar optical properties has to be integrated in a device where the radiative emission within the 8–13 μm wavelength region has to be maximized and the absorbed power from the incoming atmospheric radiation, non-radiative contributions, and absorption of solar power must be minimized. All these requirements can be satisfied acting on the different constituents of the cooling device, in particular tailoring optical properties of engineered metamaterial.

The global AM1.5 solar spectrum in the Mediterranean area has an irradiance value of about 1000 W/m^2 and a significant absorption of this power flux can easily diminish the cooling effect of the device. For this reason, the passive radiative coolers reported in the literature are utilized in the night. For a daytime application, it is important that solar incoming radiation is highly reflected (95%) in all spectral wavelength range, except for the 8–13 micron window (770–1250 cm^{-1}), where a strong selective emission is required. In case of a lack of selectivity in such IR region, a significant absorption of the atmospheric radiation outside the transparency window can occur, not allowing the cooling device to achieve temperatures significantly below the ambient temperature.

Moreover, certain atmospheric conditions can significantly influence the cooling potential, reducing the transmittance window. Particularly, the parameter mostly affecting the cooling potential is the atmospheric humidity: in a cloudy sky, the atmosphere can be completely opaque to IR radiation, invalidating any beneficial effect of the cooling device.

Within the present work, a demonstrative experiment has been performed on a proprietary prototypal device, as proof of concept of daytime passive radiative cooler properties of engineered metamaterials that exposed to direct solar irradiation, can stay well below ambient temperature.

The engineered metamaterial has been fabricated by means of a low cost, scalable sputtering process starting from cheap and abundant materials

For achieving high emissivity, silicon oxides, and nitrides have been included in the chemical composition of the materials, fabricated in form of thin films, and characterized in the full wavelength range, to validate the simulations in a region where commercial database have a lack of data. Moreover, the choice of amorphous oxides and nitrides of a same element (e.g., silicon oxide and nitrides, aluminum oxide and nitrides), was also based on the quantum tunneling of phonon dispersion heat transfer in the Heisenberg picture (the closest picture of condensed matter physics to the classical macroscopic physics [34]), where vibrational excitations can tunnel, quantum-like, from one lattice to another similar, if they are positioned at nanometer distance (e.g., in contiguous layers of the metamaterial).

Clearly, once defined the desired sequence of layers, it is necessary to optimize a sequential sputtering process. It typically suffers from modifications of ideal layer deposition conditions (e.g., substrate temperature increase) that can modify selected ideal process parameters. Moreover, chemical compatibility of different materials in contiguous layers (as coming from optical simulation and heat transfer considerations) is the real challenge of metamaterial realization. For example, plasma oxygen can be dangerous for the silver layer, so that contiguous layers cannot be oxides. Reticular mismatch from a layer to another can cause tensile or compressive stresses that compromise adhesion, or metamaterial mechanical stability. The process optimization is specific for every substrate and for every thickness of each single layer constituting the coating. Furthermore the process optimization is different depending on the thickness of the layers (thin (>100 nm) and ultrathin (<15 nm) layers), which do not have bulk properties as those simulated. As an example, in the present work, based on mechanical and thermal considerations, a metallic substrate was selected to build the device, but a silicon substrate (as in the work of Raman [24]) would have provided a better adhesion between the materials constituting the coating. In this regard, a better adhesion on aluminum will be the objective of a future revised prototype. Most probably the first adaptive layer is too thin (only 10 nm), causing mechanical stresses due to mismatching of reticular structures. A way to overcome this problem is to heat the substrate before layers deposition. Another potential cause of delamination is the formation of ammonium ions; which, in form of gas, bubble out from the metamaterial, since in the selected testing conditions AlN can hydrolyze. Similar consideration can be applied to Tungsten adaptive layer, which could be oxidized, forming tungsten oxides (explaining the pink color of delaminated parts).

Concerning the device, as well described in the previous section, a greenhouse effect inside the prototype case was detected, due to the absence of proper lateral sunshades. Despite such initial difficulty, the CM plate has demonstrated to be able to reach a temperature well below the ambient one, showing reproducible behavior also when tested in comparison with its only substrate.

Based on the evidences acquired from this proof of concept, the future work will be devoted to optimize the prototype, to newly test it using the same promising engineered cool metamaterial, and to apply it as platform for material testing on the field. As a future step, the system will be equipped with lateral sunshades and subsequently tested under direct sun irradiation. At the same time, considering that CMs are very interesting materials also when exposed directly to air, additional work will be devoted to optimize formulations for materials to be directly used in radiative roofing, further improving their optical properties and facing the issue of adhesion on different substrates and performance stability in air.

In summary, within the present work, the development and testing of a new prototype of daytime radiative cooler device, that can be used to mirror solar radiation and to emit heat in dark cool sky through atmospheric window, was performed, achieving promising results in terms of cooling efficacy (−12 °C registered under the experimental conditions tested). A cross-disciplinary approach in the development of CM and device is expected to be the key factor in view of the deployment of this promising technology.

5. Conclusions

Within the present work, a prototype of daytime passive radiative cooler has been developed and tested during daytime, starting from an engineered metamaterial formulation (named CM) by means of photonic approach method combined to a low cost scalable sputtering deposition process, passing through the manufacturing of a rectangular box device. Promising results have been obtained for the CM surface, which has reached a steady state temperature well below the ambient temperature (−12 °C) under direct sunlight. The whole process of design, fabrication, characterization of the CM and the testing on field of the device are described. However, some design limitation are identified, both for the CM and for the device, such as coating stability on some substrates (e.g., aluminum) and the absence of sunshades on the lateral wall of the device.

The obtained experimental evidences will be fundamental in driving the future research activities for the development of new metamaterials and new devices, following a cross disciplinary approach at material/system level.

6. Patents

Starting from the results here reported work is in progress to rethink a metamaterial formulation able to control electromagnetic and optic waves in order to radiate excess of heat out of the space, being at the same time stable on different construction substrates as a smart tailorable skin.

Author Contributions: A.C.: conceptualization, methodology, investigation, data analysis, original draft preparation, writing–reviewing–editing. G.V.: prototype project and development, investigation, figure and photos curation. E.G.: optical filter project definition, sputtering process optimization, data analysis, writing. M.F.: conceptualization, semi-empirical optical filter simulation, sputtering deposition on a pre-industrial plant, data analysis, writing. M.L.: funding acquisition, supervision, writing–reviewing. M.Z. investigation, writing–rewriting, supervision. All authors agreed to the published version of the manuscript.

Funding: This research was funded by MISE, Italian Ministry of Economic Development within the Programme 2019-2021, "Ricerca di Sistema Elettrico".

Acknowledgments: In this section, we would like to acknowledge Alessandro Antonaia for introducing us to this research topic and helping us with Figure 3, before his retirement, and Angelo Merola for thermocamera images acquisition.

Conflicts of Interest: The authors declare no conflict of interest.

References

1. Pachauri, R.K.; Allen, M.R.; Barros, V.R.; Broome, J.; Cramer, W.; Christ, R.; Church, J.A.; Clarke, L.; Dahe, Q.; Dasgupta, P.; et al. *Climate Change 2014: Synthesis Report. Contribution of Working Groups I, II and III to the Fifth Assessment Report of the Intergovernmental Panel on Climate Change*; Pachauri, R.K., Meyer, L.A., Eds.; IPCC. Geneva, Switzerland, 2014; p. 151. Available online: https://www.ipcc.ch/report/ar5/syr/ (accessed on 12 August 2020).

2. Rosenzweig, C.; Solecki, W.; Hammer, S.A.; Mehrotra, S. Cities lead the way in climate change action. *Nature* **2010**, *11*, 467–909. [CrossRef] [PubMed]

3. Arnfield, A.J. Two decades of urban climate research: A review of turbulence, exchanges of energy and water, and the urban heat island. *Int. J. Climatol.* **2003**, *23*, 1–26. [CrossRef]

4. Oke, T.R. City size and the urban heat island. *Atmos. Environ.* **1973**, *7*, 769–779. [CrossRef]

5. EIA Report 2019: International Energy Outlook 2019 with Projections to 2050 U.S. Energy Information Administration (EIA). Available online: https://www.eia.gov/ieo (accessed on 12 August 2020).

6. Wenz, L.; Levermanna, A.; Auffhammere, M. North–south polarization of European electricity consumption under future warming. *Proc. Natl. Acad. Sci. USA* **2017**, *114*, E7910–E7918. [CrossRef] [PubMed]

7. Santamouris, M.; Kolokotsa, D. On the impact of urban overheating and extreme climatic conditions on housing, energy, comfort and environmental quality of vulnerable population in Europe. *Energy Build.* **2015**, *98*, 125–133. [CrossRef]

8. Santamouris, M. Cooling the cities—A review of reflective and green roof mitigation technologies to fight heat island and improve comfort in urban environments. *Solar Energy* **2014**, *103*, 682–703. [CrossRef]

9. O'Malley, C.; Piroozfar, P.; Farr, E.R.; Pomponi, F. Urban Heat Island (UHI) mitigating strategies: A case-based comparative analysis. *Sustain. Cities Soc.* **2015**, *19*, 222–235. [CrossRef]

10. Konopacki, S.; Akbari, H.; Pomerantz, M.; Gabersek, S.; Gartland, L. *Cooling Energy Savings Potential of Light-Colored Roofs for Residential and Commercial Buildings in 11 U.S. Metropolitan Areas*; Lawrence Berkeley National Laboratory: Berkeley, CA, USA, 1997. Available online: https://www.epa.gov/sites/production/files/201408/documents/coolingenergysavingspotentialoflightcoloredroofs.pdf (accessed on 12 August 2020).

11. Baniassadi, A.; Sailor, D.J.; Crank, P.J.; Ban-Weiss, G.A. Direct and indirect effects of high-albedo roofs on energy consumption and thermal comfort of residential buildings. *Energy Build.* **2018**, *178*, 71–83. [CrossRef]

12. Levinson, R.; Berdahl, P.; Akbari, H. Solar spectral optical properties of pigments—Part I: Model for deriving scattering and absorption coefficients from transmittance and reflectance measurements. *Sol. Energy Mater. Sol. Cells* **2005**, *89*, 319–349. [CrossRef]

13. Berdahl, P.; Chen, S.S.; Destaillats, H.; Kirchstetter, T.W.; Levinson, R.; Zalich, A.M. Fluorescent cooling of objects exposed to sunlight—The ruby example. *Sol. Energy Mater. Sol. Cells* **2016**, *157*, 312–317. [CrossRef]

14. Karlessi, T.; Santamouris, M.; Apostolakis, K.; Synnefa, A.; Livada, I. Development and testing of thermochromic coatings for buildings and urban structures. *Sol. Energy* **2019**, *83*, 538–551. [CrossRef]

15. Baniassadi, A.; Sailor, D.J.; Ban-Weiss, G.A. Potential energy and climate benefits of super-cool materials as a rooftop strategy. *Urban Clim.* **2019**, *29*, 100495. [CrossRef]

16. Givoni, B. Solar heating and night radiation cooling by a roof radiation trap. *Energ. Build.* **1977**, *1*, 141–145. [CrossRef]

17. Lu, X.; Xu, P.; Wang, H.; Yanj, T.; Hou, J. Cooling potential and applications prospects of passive radiative cooling in buildings: The current state-of-the-art. *Renew. Sustain. Energy Rev.* **2016**, *65*, 1079–1097. [CrossRef]

18. Liu, C.-H.; Ay, C.; Tsai, C.-Y.; Lee, M.-T. The Application of Passive Radiative Cooling in Greenhouses. *Sustainability* **2019**, *11*, 6703. [CrossRef]

19. Hossain, M.M.; Gu, M. Radiative Cooling: Principles, Progress, and Potentials. *Adv. Sci.* **2016**, *3*, 1500360. [CrossRef] [PubMed]

20. Gentle, A.R.; Smith, G.B. "A subambient open roof surface under the mid summer sun". *Adv. Sci.* **2015**, *2*, 1500119. [CrossRef]

21. Rephaeli, E.; Raman, A.; Fan, S. Ultra broadband photonic structures to achieve high-performance daytime radiative cooling. *Nano Lett.* **2013**, *13*, 1457–1461. [CrossRef]

22. Pendry, J.B.; Schurig, D.; Smith, D.R. Controlling Electromagnetic fields. *Science* **2006**, *312*, 1780–1782. [CrossRef]

23. Liu, Y.; Zhang, X. Metamaterials: A new frontier of science and technology. *Chem. Soc. Rev* **2011**, *5*, 2494–2507. [CrossRef]

24. Hossain, M.M.; Jia, B.; Gu, M. A Metamaterial Emitter for Highly Efficient Radiative Cooling. *Adv. Opt. Mater.* **2015**. [CrossRef]

25. Ko, B.; Lee, D.; Badloe, T.; Rho, J. Metamaterial-Based Radiative Cooling: Towards Energy-Free All-Day Cooling. *Energies* **2019**, *12*, 89. [CrossRef]

26. Fu, Y.; Yang, J.; Su, Y.; Du, W.; Ma, Y. Daytime passive radiative cooler using porous alumina. *Sol. Energy Mater. Sol. Cells* **2019**, *191*, 50–54. [CrossRef]

27. Lim, X.Z. The Supercool Materials That Send Heat to Space. *Nature* **2020**, *577*, 18–20. [CrossRef]

28. Kong, A.; Cai, B.; Shi, P.; Yuan, X.C. Ultra-broadband all-dielectric metamaterial thermal emitter for passive radiative cooling. *Opt. Express.* **2019**, *27*, 30102–30115. [CrossRef]

29. Zhai, Y.; Ma, Y.; David, S.N. Scalable-manufactured randomized glass-polymer hybrid metamaterial for daytime radiative cooling. *Science* **2017**, *355*, 1062–1066. [CrossRef]

30. Esposito, S.; D'Angelo, A.; Antonaia, A.; Castaldo, A.; Ferrara, M.; Addonizio, M.L.; Guglielmo, A. Optimization procedure and fabrication of highly efficient and thermally stable solar coating for receiver operating at high temperature. *Sol. Energy Mater. Sol. Cells* **2016**, *157*, 429–437. [CrossRef]

31. Castaldo, A.; Ferrara, M.; Antonaia, A. Low emissive sputtered coating for smart glazing. In Proceedings of the IEEE: 18th Italian National Conference on Photonic Technologies (Fotonica 2016), Rome, Italy, 6–8 June 2016; IET: London, UK, 2016; pp. 1–4. [CrossRef]

32. Antonaia, A.; Addonizio, M.L.; Esposito, S.; Ferrara, M.; Castaldo, A.; Guglielmo, A.; D'Angelo, A. Adhesion and structural stability enhancement for Ag layers deposited on steel in selective solar coatings technology. *Surf. Coat. Technol.* **2014**, *255*, 96–101. [CrossRef]

33. Ferrara, M.; Castaldo, A.; Esposito, S.; D'Angelo, A.; Guglielmo, A.; Antonaia, A. AlN–Ag based low-emission sputtered coatings for high visible transmittance window. *Surf. Coat. Technol.* **2016**, *295*, 2–7. [CrossRef]

34. Chiloyan, V.; Garg, J.; Esfarjani, K.; Chen, G. Transition from near-field thermal radiation to phonon heat conduction at sub-nanometre gaps. *Nat. Commun.* **2015**, *6*, 6755. [CrossRef]

Article

The Microclimate Design Process in Current African Development: The UEM Campus in Maputo, Mozambique

Giovanni M. Chiri [1,*], **Maddalena Achenza** [1], **Anselmo Canì** [2], **Leonardo Neves** [2], **Luca Tendas** [1] **and Simone Ferrari** [1,*]

1   Dipartimento di Ingegneria Civile, Ambientale e Architettura (DICAAR), University of Cagliari, 09123 Cagliari, Italy; maddalena.achenza@unica.it (M.A.); lucatendas@yahoo.it (L.T.)
2   Faculty of Architecture and Physical Planning, University Eduardo Mondlane, Maputo 1102, Mozambique; cani@africamail.com (A.C.); leonardo.mario@outlook.pt (L.N.)
*   Correspondence: g.chiri@unica.it (G.M.C.); ferraris@unica.it (S.F.); Tel.: +39-706755634 (G.M.C.); +39-706755307 (S.F.)

Received: 3 March 2020; Accepted: 30 April 2020; Published: 7 May 2020

Abstract: Even if current action towards sustainability in architecture mainly concerns single buildings, the responsibility of the urban shape on local microclimate has largely been ascertained. In fact, it heavily affects the energy performances of the buildings and their environmental behaviour. This produces the necessity to broaden the field of intervention toward the urban scale, involving in the process different disciplines, from architecture to fluid dynamics and physics. Following these ideas, the Masterplan for the Campus of the University Eduardo Mondlane in Maputo (Mozambique) develops a methodology that integrates microclimatic data and analyses from the initial design model. The already validated software ENVI-met (Version 4.1, ENVI_MET GmbH, Essen, Germany) acts as a useful 'feedback' tool that is able to assess the microclimatic behaviour of the design concept, also in terms of outdoor comfort. In particular, the analysis focused on the microclimatic performances of a 'C' block typology east oriented in relation to the existing buildings, in Maputo's specific climatic characteristics. The initial urban proposal was gradually evaluated and modified in relation to the main critical aspects highlighted by the microclimatic analyses, in a sort of circular process that ended with a proposed solution ensuring better outdoor comfort than the existing buildings, and which provided an acceptable balance between spatial and climatic instances.

Keywords: urban form; urban microclimate design; city; sustainability; sustainable development

---

1. Introduction

The centrality of the city with reference to energy efficiency has been confirmed by several scientific contributions and in various disciplines. Nevertheless, in current approaches, the majority of legislative guidelines and design activities are related to individual buildings. Interesting studies (De Pascali, 2008 [1]) have shown how the long-term benefits obtained by the application of energy saving policies to single buildings tend to reduce their effectiveness. Thus, it is strongly necessary to shift the focus from the city, conceived as a collection of single buildings, to the city as an organism with its own shape and its own climatic behaviour: "A significant share of energy is, in fact, linked to the physical and functional relationships which are established, or may be established, among the various elements of the settlement, and which determine the city's shape and organization" [1]. The support of urban microclimate design in the reconfiguration and re-organization of the urban form could lead urban morphology towards more efficient configurations. Through this process, we will be able to reduce *neg*-entropic requirements, raising both the quality and liveability of the spaces of the cities.

Energy Balance in the Urban Microclimate Design

As stated by Balaras et al. (2019) [2], the employment of urban microclimate design can transform buildings and the built environment in cities from the cause of issues into one of the solutions to today's economic, environmental and social challenges. From this point of view, the focus is usually more on the reduction of energy consumption, generated by artificial heating and cooling, and on the use of renewable energies (see, e.g., Susowake et al., 2019 [3], Sánchez Cordero et al., 2019 [4], Mancini and Lo Basso, 2020 [5]) rather than on the capability of urban environments to efficiently use energy. As a matter of fact, there is a relationship between the energy balance and the urban form that can be summarized in Equation (1), proposed, among others, by M. Santamouris (2001) [6] and De Pascali (2008) [1]:

$$Q_R + Q_T = Q_E + Q_L + Q_S + Q_A \tag{1}$$

where it is highlighted that the energy gain of the city $Q_R + Q_T$ (where Q_R is the net radiative flux, directly influenced by the urban surface radiation, and Q_T is the anthropogenic heat flux, given by transportation, heating systems, etc.) contrasts with the heat flow that the system stores and the exchange inside $Q_E + Q_L + Q_S + Q_A$ (where Q_E is the sensible heat, Q_L the latent heat, Q_S the stored energy and Q_A the energy transferred to/from the system through advection), clearly pointing out the roles of urban form in the energy performance of the settlement. Following, among the others, the seminal work of T.R. Oke (1982) [7], one of the key parameters in urban microclimate design is the space among buildings (urban canyons), which plays a significant role in the efficient functioning of urban environments (see, for instance, the studies of Chiri e Giovagnorio, 2015 [8] and Giovagnorio et al., 2017 [9]). In particular, the ratio of the building height (H) to the distance between building façades (W) affects the air flows (see, for instance, Badas et al., 2020 [10], Garau et al., 2019 [11], Badas et al., 2019 [12], Garau et al., 2018 [13], Badas et al., 2017 [14], Ferrari et al., 2017 [15]) and, consequently, the natural ventilation potential and the capability to disperse or accumulate pollutants, humidity and heat, both outside and inside the buildings (Costanzo et al., 2019 [16]). As a consequence, *H/W* plays a relevant role on the Urban Heat Island (UHI) phenomenon and, in turns, in the urban environment energy balance. For instance, as shown by NG (2010) [17], when the wind is perpendicular to the fabric orientation, a *H/W* ratio larger than 2 drastically reduces air penetration into the urban canyons, almost cancelling the natural ventilation at pedestrian level. Moreover, the *H/W* ratio bias the solar radiation penetration: Santamouris (2001) [6] suggests to designers, as a good construction technique, to increase/decrease the *H/W* ratio when the latitude decreases/increases, in order to reduce/enhance the light penetration and, as a consequence, the temperature, reducing the use of air conditioning or heating. In this way, the urban microclimate design contributes to the energy efficiency of the cities and to the reduction of the UHI phenomenon, as well as to the optimization of the outdoor human comfort. Moreover, the *H/W* ratio plays a relevant role in the control of outdoor human comfort and thermal stress. Outdoor human comfort is one of the key parameters for quantifying the results of an intervention based on urban microclimate design, but it is difficult to define in one single quantity, as both objective quantities (such as, for instance, humidity, temperature, air velocity, solar exposition, etc.) and subjective quantities (such as, for instance, human activities, heat sensitivity, clothing, etc.) contribute to its definition. An extensive classification can be found in the review of Coccolo et al., 2016 [18], where the outdoor human comfort indices have been classified into three main categories, namely the thermal indices, the empirical indices and the indices based on Linear Equations. Among the thermal indices, the Predicted Mean Vote (PMV), based on the energy balance of a human being, is an index representing the average temperature (and the related thermal stress) sensed by a group of people. PMV was firstly introduced by Fanger (1970) [19] for the quantification of indoor human comfort, but was then adapted to outdoor human comfort through the use of micrometeorological variables (such as wind speed, relative humidity, mean radiant temperature, solar irradiation and outdoor temperature) and adding, among other things, human activity and clothing factors (Coccolo et al., 2016 [18]). Even if those parameters can be measured at some points through sensors, only the adoption of microclimatic software makes it possible to study the performance of the built environment during the design phase over the entire investigation field. Among available microclimatic software, ENVI-met (Version 4.4, ENVI_MET GmbH, Essen, Germany) Bruse and Fleer, 1998 [20], https://www.envi-met.com/ [21]), employing the German VDI 3787 Standard, 2008, to compute the PMV, has often been used to simulate outdoor comfort in the recent past (e.g., Abdi et al., 2020 [22]; Ouali et al., 2020 [23]; Hassan et al., 2020 [24]; Limona et al., 2019 [25]; Shinzato et al., 2019 [26]; Karakounos et al., 2018 [27]; Nasrollahi et al., 2017 [28]; Ghaffarianhoseini et al., 2015 [29]). As reported by Fabbri and Costanzo, 2020 [30], ENVI-met is the most widely used software for simulating the urban outdoor microclimate; they reported a list of 59 recent (from 2013 to 2019) scientific papers where ENVI-met was employed to simulate the outdoor microclimate. Among these, 19 showed simulations that had not been validated with field-measured data, mainly with the target of comparing different planned solutions, similarly to what has been done in the present case (i.e., the comparison of the microclimatic performances of present building to the planned ones). This is because ENVI-met includes the main variables and processes involved in urban microclimate evolution, such as turbulence, heat, pollutant transport and dispersion, radiation fluxes within buildings and natural/artificial soils, influence of vegetation, etc., and, differently from most of the other computational tools, it is able to do this while at the same time reproducing the daily variations in sun position and its influence on shading/insulation. So ENVI-met includes in a single piece of software the numerical simulation of fluid dynamics, thermodynamics and radiation balance in complex urban domains. Therefore, despite some inherent limitations due to the discretization of the studied domain and to the approximations introduced by the numerical solution of complex equations typical of this kind of software, this tool has been demonstrated to produce valuable and useful feedback, through which a designer can direct urban design so as to improve its energy and microclimatic performance. In the present paper, the targets are:

1. to evaluate whether or not the planned solution can have better microclimatic performances, also in terms of outdoor comfort, than the existing buildings;
2. to test a recursive and multidisciplinary design process, where different disciplines such as Architecture and Fluid Dynamics are involved and together contribute to the proposed solutions;
3. to apply this process to a social and cultural context, the African city of Maputo (Mozambique), where the urban microclimate design is not usually applied.

2. The Case Study: History, Future Perspectives and Critical Issues

The case study is in the preliminary stages of the urban design process concerning the development of the Eduardo Mondlane University Campus in Maputo (Mozambique).

The Eduardo Mondlane University (Universidade Eduardo Mondlane in Portuguese; UEM) is the oldest and largest university in Mozambique and is located in the Sommerschield district (Figure 1), right in the interface between the formal city, served by urban infrastructure and services, and the spontaneous informal city, lacking almost any type of service. It occupies a peripheral position in relation to the formal colonial city, located to the south, and runs alongside one of its most important access routes, the Avenida Julius Nyerere, which favours the connection of the city to the north of the country. The whole area occupies around 147,000 square metres, and it hosts several faculties of the Universidade Eduardo Mondlane: one high school, some administrative pavilions, a library, a scientific research centre, a canteen and residences for students and professors. The Campus area is used not only by people attending the University, but also by the rest of the population. In fact, it houses several public functions, including a bank, restaurants and sports activities. UEM may be able to deal with the social, cultural and sports development of the society in general, addressing the whole community. In fact, there has been an increase in demand among the population for cultural spaces that could ease social inclusion and sharing. A possible integration of cultural spaces within the Campus in the educational and tourist circuit of Maputo is therefore a fundamental aspect with great potential which, accompanied by a good promotion strategy, would contribute to the promotion not only of the University but, above all, of the African cultural identity.

Figure 1. The Sommerschield district in 2017. The UEM Campus is highlighted in the centre of the area. Behind the north boundary there is the informal development.

The Campus currently presents several problems related to a fragmented and chaotic structure, to the lack of infrastructure and internal order, which do not allow full liveability, orientation and movement. Moreover, this does not facilitate its opening towards the city, marking the break between the autochthonous informal city and the formal colonial one. The main current issues, emerging on the basis of the analysis of student reports, are mainly related to spatial organization, quality of learning spaces, quality of public external spaces, climatic comfort (both inside and outside the building), and connection with the city. These interviews highlight critical points of the Campus in its current condition, but also constitute a good starting point for identifying important topics for the design of infrastructure development. In terms of spatiality, the Campus layout, even in an attempt to recall the orthogonal "colonial" mesh, lacks an internal organization, and the various buildings appear disconnected from one another. The extension of the Campus and its organization require an increase in the equipment of public spaces and should be intended for pedestrian flow.

The 2018 UEM Rector report, taking up the 2018–2028 Strategic Plan, lists some possible transformation scenarios aimed in these directions, such as the construction of buildings for the School of Art, the increase in the number of dormitories for students, the creation of deposits for the historical archives of Mozambique, the recovery of spaces for sport, a project for a health centre, the expansion of the structures of some departments, the increase in services and multipurpose cultural spaces, which would also allow significant financial income for the Campus.

Concerning the microclimate, the situation is not optimal. The climate characteristics associated with a not fully sustainability-oriented design dramatically affect both the liveability of the internal spaces and the outdoor comfort. In fact, the interior spaces present serious difficulties in obtaining good thermal comfort without the use of air conditioning equipment, and the outdoor space is not enjoyable during the day.

The microclimate on the campus is tropical, with high temperatures associated with a high relative air humidity. This condition makes control of the local microclimate difficult to attain, especially if the architectural solutions do not consider the problem of orientation and protection of the rooms against the direct incidence of sunlight as the predominant design principle. On the other hand, in the dry season, temperatures can drop significantly, and the hydrometric degree can also be uncomfortably low.

The Process for the Design of the Campus

The Sommerschield district, in which the UEM is located, remained out of planning until 1965. The UEM Campus was initially organized in a project by PROFABRIL, a Portuguese company for industrial projects. This plan, on which the first buildings of the Campus were built, established general distribution rules for activities and functions through their zoning. It was only in the early 2000s that the increasing student population, largely determined by a strong economy growth, made it necessary to reconsider the general layout of the campus in order to provide better accommodation for both students and professors.

The first proposed Masterplan of the Campus dates to 2004. It was designed by a team of local architects lead by José Forjaz, an architect of Portuguese origins, who, since 1978, represented a prestigious figure in both political and cultural fields. Forjaz was director of the Faculty of Architecture of UEM and director of the DNH (Direcção Nacional de Habitação) of the Ministry of Public Works and Housing. Thanks to the efforts of his research group, even if not implemented, the 2004 plan still constitutes a point of reference for the university, and provides a useful key for the analysis and understanding of the context.

A critical reading of the plan in fact provides interesting support for the new project development. The situation in 2004, before the Forjaz's Masterplan, shows a disordered structure in which some buildings follow the north-south Cartesian orientation, others are oriented according to the distribution axis of the formal city and others follow their own logic. The buildings are concentrated in different areas of the campus, taking up the concept of "subdivision by zones and functions" of the PROFABRIL plan.

Looking at Figure 2, the first panel (Figure 2a, phase 0), certifies that in 2004 the area for sport fields, student and professor dormitories, the university library, the faculties of Science, Economics, Letters, Agronomy, Biology, Laboratories, the botanical garden building, already existed. Technical rooms, a sports pavilion, an area for the historical archive and one occupied by TELEVISA, and a private radio and television group are also present. The distribution of the road system for vehicles and parking areas is also reported. The second panel (Figure 2b, phase 1), shows the interventions deemed immediate by Forjaz's Plan. The third panel shows Forjaz's plan in its whole (Figure 2c, final phase). The proposal includes the distribution of the buildings along the north-south axis and aims to cover the entire available area, concentrating in the eastern part. The idea is to adopt a rectangular module, intended to house the classrooms, which is repeated vertically along a narrower 'slat', or a corridor intended for connection. Each classroom is combined with a patio and, at a less dense rhythm, by three smaller square modules, intended for the auditorium, while the services are arranged in smaller modules, mirroring the classrooms.

The external connection is solved through a road that, starting from the one already present in the centre of the area, runs parallel to the Campus border from north-east to south, and then continues in the different areas, joining the Rua dos Presidentes, Avenida De França and Avenida Vladimir Lenine, roads that connect the Campus to the city. The ramifications towards the centre of the Campus are therefore eliminated, as well as the parking areas, which are instead distributed in various points along the road that surrounds the main area. The road connecting Avenida Julius Nyerere has been moved, almost cutting the Botanical Garden in half. The pedestrian traffic is thought through covered walkways, which constitute the corridors on which the classrooms are grafted.

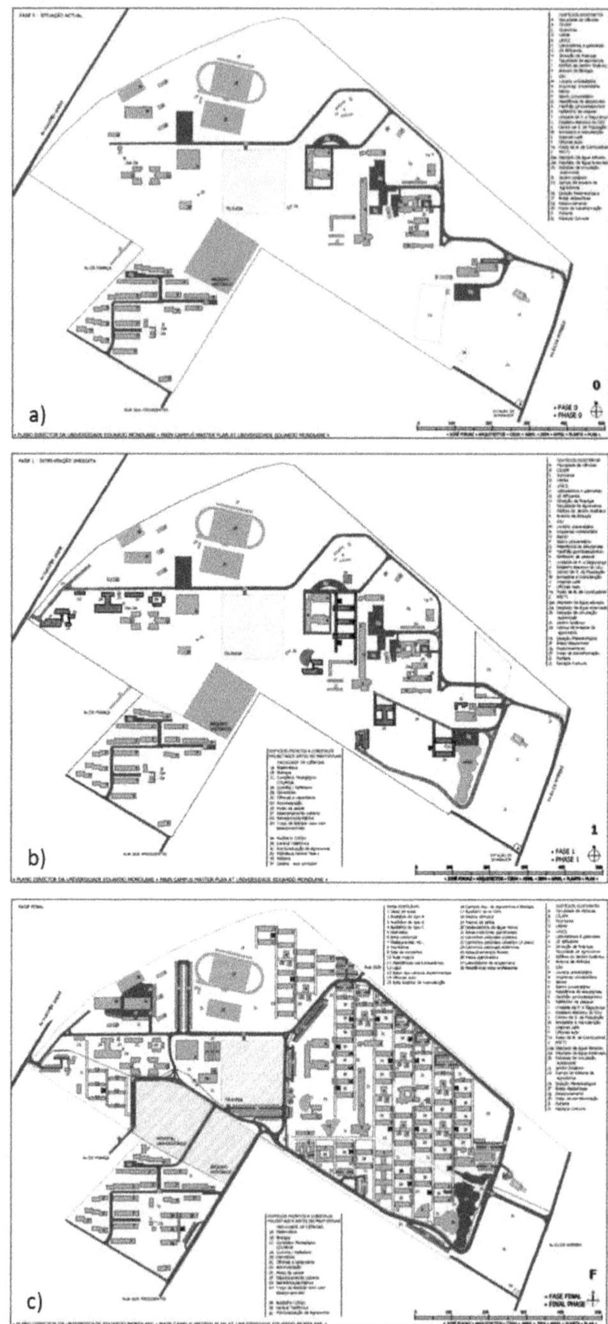

Figure 2. The Campus in 2004. (**a**) starting situation (phase 0): the Campus in 2004, (**b**) the interventions deemed immediate by the Forjaz's Plan (phase 1), (**c**) Forjaz's Masterplan for 2006–2020 (final phase).

Overlapping Forjaz's Plan with the current state, it is easily concluded that the evolution of the conformation of the Campus did not follow the direction given by the 2004 Masterplan. The buildings built after 2004 that incorporate the plan are those that were highlighted in Phase 1 (Figure 2a) with the exception of the lake. The current Department of Mathematics and Computer Science (Figure 3) is the only building to resume the setting for "grafts" that characterizes the Plan. The structure is made up of two almost identical rectangular blocks that graft onto a covered corridor. However, this body is interrupted by a third rectangular block, as wide as the others but twice as long. The reasons for which Forjaz's Plan was abandoned are unknown but, as can be seen from the current state of affairs, the Campus is still incomplete and lacking in the facilities and services that should have been provided for the various departments. In Figure 4, the current situation of the Campus is shown; among the issues of the present Campus, three important points have to be highlighted:

1. The central area is still occupied by Televisa. The telecommunications company established its headquarters inside the perimeter of the Campus from the time of the civil war, occupying an extremely strategic position. This area is scarcely exploited, since it is an area with few buildings and a lot of free space.
2. The south-west belt is occupied by the presence of a large construction site owned by the Chinese Government. There is currently a pavilion and a new entrance to the campus.
3. The southern part of the Campus is bordered by a vast area currently free of buildings whose ownership and intended use are unknown. This area could play an important role and contribute to the redevelopment of UEM.

Figure 3. Existing modular blocks. The Department of Mathematics and Computer Science is the only building to resume the setting for "grafts" that characterizes the Plan.

Figure 4. Current situation of the Campus. The four main entrances are signed with an arrow. In the north, the informal district with the typical steel-covered roof. On the right the coast with the line of recent skyscrapers. The green area is the botanical garden.

3. Materials and Methods

3.1. Proposal

The case study presented herein can be described as an iterative and multidisciplinary process of design and verification of microclimatic performance in relation to the local conditions of the site. The work, therefore, aims at defining a repeatable procedure, which integrates the principles of architectural design with the urban microclimate design into a single process in order to optimize the energy performance of the new urban fabric, to improve its comfort levels and to reduce construction costs, all of which are necessary to achieve the highest standards. The project aims at establishing a permanent research board on the topic of sustainable urban design focused, in particular, on the development of UEM Campus. In terms of the architectural design approach, the new proposal does not differ significantly from the previous one except for the proportion of the blocks, whose typology has been slightly modified in order to fit with current spatial needs.

The desire to maintain the general layout of Forjaz's proposal largely originates in reducing the urbanization costs, but also in the recognition of a rational urban design, the good balance between public and semi-public spaces, and the clear reference to the formal grid layout of the colonial city, which became an element of order in relation to the chaotic and informal development of spontaneous surrounding neighbourhoods. Some optimization has been adopted for the public spaces, roads, sidewalks and multipurpose public building. The comparison between Forjaz's proposal of a modular building typology, the currently adopted one, and the one planned by the 'País Project' reveals the evolution of the global design. Even if the collected documentation does not provide information regarding the elevation envisaged in Forjaz's Plan, it sounds reasonable to consider an elevation of two or three levels at least. The rectangular volumes are intended for the main functions and are inserted into the longitudinal connection element. The services, placed independently, are separated from the main volumes and are repeated at irregular intervals along the 'slat'. In the current situation, the only building that appears to resume Forjaz's Plan is located in the central part of the Campus. The structure follows the orientation and the proposed arrangement, but the model has undergone many changes. The rectangular volumes are wider and go directly into the 'walkway batten'. The functional program of each block is distributed over three levels, two of which are intended to host classrooms. The complementary volumes for services remain, while the square volume in the centre of the patio becomes a loggia, covered on the top floor. The staircase runs parallel to the 'slat', and two staircases have been added on the side of each rectangular volume, supported by a septum, each one inconveniently serving a different floor. Each volume has a different height, and this is mainly at disadvantage of the connecting batten, now difficult to perceive because its shape appears fragmented, representing just a covered walkway. In the new proposal the type of the block is made up of three volumes that are grafted onto a narrower perpendicular volume. Even if the proposal maintains the position and the general setting of Forjaz's Plan, the spaces and connections are optimized and based on the internal functional program. The project follows a specific hierarchy, in which the corridor acquires character, becoming a liveable and multifunctional space that allows distribution in the various branches, each with a specific function. The rectangular modules are arranged along the 'slat', marking a 1:2.7 ratio with the courtyard; the serving rooms are rethought and acquire meaning, the general design appears clearer and more orderly. Even if the 'comb-shaped' south-east oriented disposition is common to the three plans, they diverge in terms of proportion of the modular blocks (Figure 5). Forjaz's 'C-shaped' block is the smallest in terms of both width and length; the courtyard is smaller even than the existing ones, and it seems to be insufficient for ensuring air circulation. The existing building has the same orientation, but different proportions, from the proposed one, which had a bigger courtyard with a similar ratio but 30% deeper. The comparison among the proposal for the typology of the modular building, the currently adopted one, and the one planned by the 'País project' reveals the evolution of the design. The proportions of Forjaz's 'C-shaped' block, of the existing 'C-shaped' block, and of the proposed País' 'C-shaped' block are listed in Table 1 and shown in Figure 5.

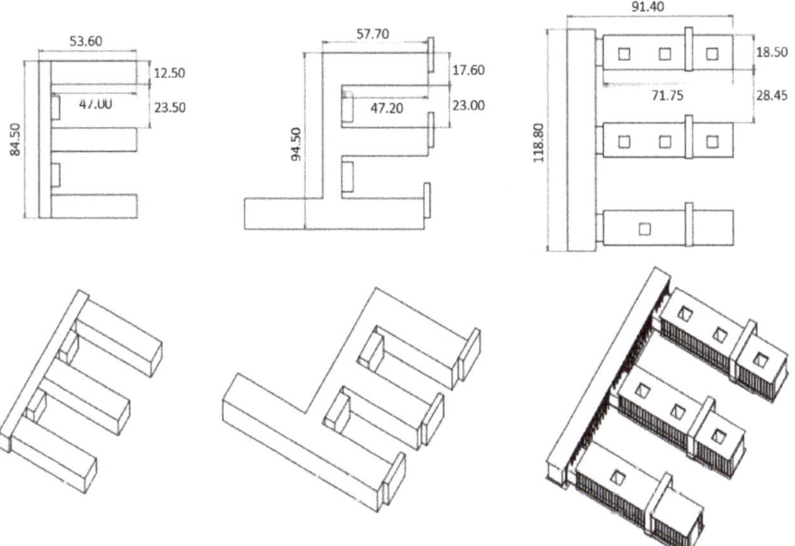

Figure 5. Different settings for the modular blocks. From left to right: Forjaz's 'C-shaped' block, existing 'C-shaped' block, and proposed País' 'C-shaped' block (measures in metres).

Table 1. Proportions of Forjaz's 'C-shaped' block, of the existing 'C-shaped' block, and of the proposed País' 'C-shaped' block.

Proportions of C-Shaped Block	Building Length [m]	Building Width [m]	Block Length (Including the Courtyard) [m]	Block Width [m]	Height [m]	Courtyard Length [m]	Courtyard Width [m]	Courtyard Ratio [-]	H/W Ratio [-]
Forjaz	47.00	12.50	84.50	53.60	7.00	47.00	23.50	1:2	3.36
Existing	47.20	17.60	94.50	57.70	11.00	47.20	23.00	1:2	2.09
País	71.75	18.50	118.80	91.40	10.50	76.00	28.45	1:2.7	2.70

The depth of the País' 'C-shaped' block (Figure 6) is largely derived from the need to ensure the correct insulation for classrooms and a better quality of the public space. In terms of urban design and orientation, the new Masterplan does not differ significantly from either the Forjaz one or from the existing buildings, except the proportions of the courtyard, which evolve 23.00 × 47.00 m (ratio 1:2) to 28.45 × 76.00 m (ratio 1:2.7). Even if this change seems minimal, its effects significantly affect the overall outdoor comfort performances. The 'Project Pais' Masterplan (2019) is shown in Figure 7; the new buildings have been split into two complexes, a northern complex and a southern complex, to improve human outdoor comfort, following the results of the simulation of the present situation (see Section 3).

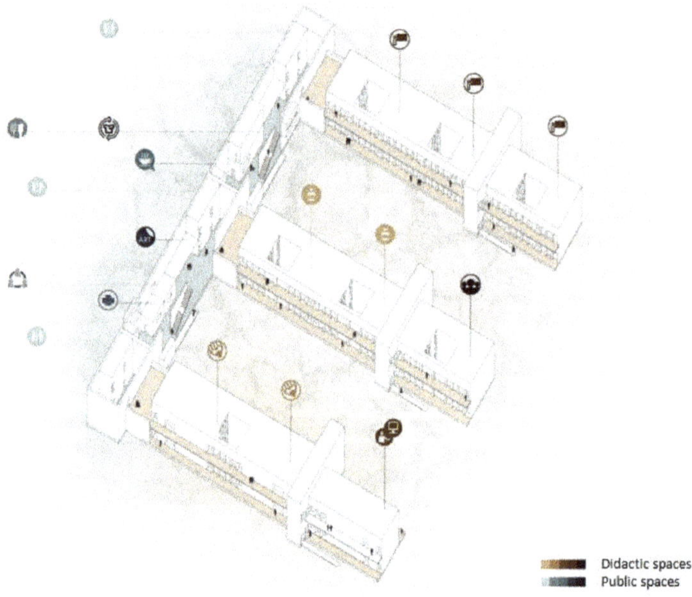

Figure 6. Typical internal distribution of modular block in 2019 proposal.

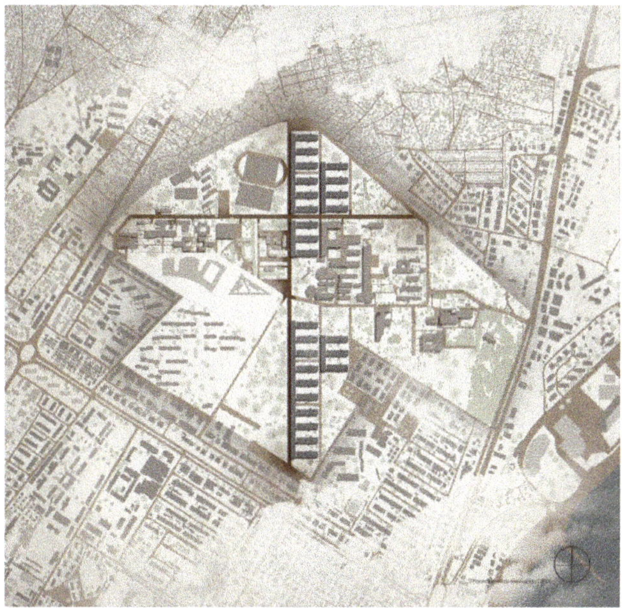

Figure 7. The Sommerschield district in 'Project Pais' Masterplan (2019); the new buildings are split into two complexes, a northern complex and a southern complex.

3.2. From Standard Planning Methodologies for Urban Microclimate Design

Even if the general design proposed by the 'Project Pais' research group represents a significant update with respect to the existing plan, the main innovation of this case study is represented by the methodology behind the process of generating the urban shape. As stated in the previous paragraphs, the current urban design does not properly take into account the microclimate behaviour of urban shape. The standard approach interprets the topic of sustainability as if it were acceptable to leave the management of the issue of thermal outdoor comfort to building technology and/or artificial air conditioning alone. Even if, as stated previously, the scientific community currently accepts that the application of energy saving policies to single building tends to reduce their effectiveness (De Pascali, 2008 [1]), the majority of legislative guidelines and design activities concern individual buildings and exclude the adoption of particular policies for the urban shape. This produces the necessity to broaden the field of intervention toward the urban scale, involving in the process different disciplines: from Architecture to Fluid dynamics and Physics. On this topic, the Masterplan for the Campus of University Eduardo Mondlane in Maputo develops a methodology on the basis of which integrate both microclimatic data and analyses start from the initial design model. The already validated software ENVI-met (see Section 1) acts as a useful 'feedback' tool that is able to verify—even qualitatively—the microclimatic behaviour of the designed concept. In this case, the analysis is focused on microclimatic consequences of a 'C' block typology southeast oriented in the climatic Maputo's specific climatic characteristics. As suggested by Blocken (2015) [31], since experimental data for the case under study were not available, the validation process can be achieved by performing generic sub-configurations contained in high-quality experimental datasets available online.

From the usual point of view of an architect (or of a planner), it is not acceptable to originate urban shape directly or automatically on the basis of microclimatic needs. It sounds reasonable, because the architecture of the city is actually the result of a complex relation between several constrains and factors, functional and even symbolic, that cannot be simplified in a sort of mechanical process of urban shape generation. Thus, this approach starts from an early layout that is gradually evaluated and modified in relation to the main critical issues highlighted by microclimatic analyses, in a sort of circular process which ends with a proposed solution that can be considered as an acceptable balance between spatial and microclimatic instances.

The proposed approach to urban microclimate design is structured through a three-step process:

- Phase 1. The present urban design is analysed from a microclimatic point of view, with the support of the already validated software ENVI-met, in order to identify primary critical situations, and test typo-morphological solutions of the project and the consequences that these choices produced on microclimatic behaviour.
- Phase 2. The initial choices of the proposed urban design are modified with the intent of improving design performance and overcoming the critical situations previously identified. The proposed modifications are studied in order to comply with certain general settings of the original project.
- Phase 3. The proposed urban design is analysed as in phase 1 to evaluate its microclimatic and outdoor comfort performances and, in case, modified and re-evaluated, until the urban design is considered satisfactory.

As mentioned in previous paragraphs, Forjaz's Masterplan has not been implemented to date. Consequently, to verify the microclimatic performance of the Forjaz's Masteplan makes no sense in terms of scientific results. Nevertheless, the current proposal largely derives from the general layout designed by the Mozambican-Portuguese architect. Even if some important improvement were set in terms of typology and spatial organization, the 'comb-shaped' layout of the new proposal is largely consistent with the previous one. In terms of macro-behaviour, the microclimatic performances of the two proposals may have a lot of similarities. While at a 'micro-scale', here intended as the scale of a single building, the two behaviours may differ significantly. Consequently, in terms of 'qualitative analysis' of the climatic performances of the Masterplan, in this case only two different configurations were evaluated: the current one and the 'Project Pais' design.

3.3. Input Data for the Simulations

As previously stated, the analyses presented in this paper were developed with the help of ENVI-met (version 4.4). ENVI-met allows the designer to evaluate the performance for specific periods of the day of many microclimatic parameters, through easy-to-understand thematic graphical maps; in particular, in this work, the focus is on wind, temperature, relative humidity and PMV.

The UEM Campus is in Maputo, which is located at around 25°58′ S, 32°35′ E directly facing the Indian Ocean, with a tropical savanna (or wet and dry) climate (Aw according to Köppen climate classification). Meteorological data regarding the town can be obtained from three meteorological stations: Maputo, Maputo MZ and Maputo/Benfica. Being in the southern hemisphere, monthly averaged air temperature usually varies from a minimum of 19.5 °C in July to a maximum of 26.9 °C in December and January, with a mean value of 23.7 °C. Relative humidity varies from a minimum of 67% in September to a maximum of 74% from December to March, with a mean value of 71%. Wind velocity varies from a maximum of 5.0 m/s on September and October to a minimum of 3.3 m/s in June, with an average of 4.3 m/s, and the prevailing (i.e., the most frequent) wind direction is south-east, that corresponds also to the dominant wind (i.e., the one with the highest velocities).

As in the tropical and subtropical climates, the most critical hours for the Urban Heat Island phenomenon are in the range 14:00–16:00 (Ferreira et al., 2012 [32]), the time chosen for the data extraction is 15:00 of the most critical day for the local situation December 21st, which is the summer solstice (in the southern hemisphere), with the simulation starting the day before. Data were saved every 30 minutes of simulated time, while the update timing was 600 s for plant processes, 30 s for surface data, 600 s for radiation/shadows and 900 s for the flow field. The input data were: a wind velocity of 4.3 m/s at an altitude of 10 m and coming from the south-east direction; a temperature of 27.0°; a relative humidity of 74%. The boundary conditions were: Simple Forcing for the meteorology; TKE (Turbulent Kinetic Energy) model (Mellor and Yamada 1982) for the turbulence model; Lateral Boundary Condition (LBC) for temperature and humidity: Forced; LBC for turbulence: Cyclic.

Regarding the spatial domain chosen for the simulations, on the left of Figure 8, the terrain elevation in metres above the mean sea level, employed in the ENVI-met simulations is shown. Moreover, we must consider that the Sommerschield district, in which the UEM Campus is located, is on the edge of the so-called 'formal city', the planned city built during the colonial age. The northern surroundings are fully covered by an 'informal' unplanned suburb built mainly of poor buildings with steel roofs. In the south-east corner there is a large botanical garden which provides humidity and shadow. The east border is close to the seafront even if some high building prevents the Campus to be full related with it. Those external parts may have a critical influence on the climatic performances of the Campus; thus, the considered domain is significantly larger than the UEM's area. In fact, the domain, taking into account the terrain altitude, is: 1020 m long, 920 m wide and 250 m high for the current situation (on the left of Figures 9–14), and 1525 m long, 1375 m wide and 250 m high for the planned situation (on the right of Figures 9–14). In both cases, individual cells are cubes of 5 m, leading to a 204 × 191 × 50-cell domain for the current situation and to 305 × 275 × 50-cell domain for the planned situation. In particular, the DTM (Digital Terrain Model) is shown, for a better visualization, in two Figures, Figures 8 and 9. In Figure 9, the soil, buildings (in black) and vegetation on the two domains (current and planned situations) are shown. The existing buildings of the Campus have a height of 11.00 m, the proposed ones 10.50 m, and the surrounding ones are between 4 and 7 m, with the relevant exception of two hotels close to the ocean, in the south-east direction, the highest building of which has a height of around 70 m. As shown by Fabbri et al. (2017) [33] and Duarte et al. (2015) [34], the vegetation plays a relevant role in the urban microclimate, so the present vegetation has been included in the simulated domain; in particular, dark green indicates very dense trees, 10m-high and with a leafless base (ID 0000T1 in the ENVI-met plant database), pale green indicates very dense trees, 15 m high, with a distinct crown layer (ID 0000SK in the ENVI-met plant database), while the buildings have the default wall parameters of ENVI-met. The terrain is "terre battue" (ID 0000TB in the ENVI-met surface database), which was chosen because, among the materials for the soil surface included in ENVI-met, it had the closest features to those of the natural soil of the Campus and its neighbourhood: very hard and dry. To improve the visualization and avoid any confusion with the data, in Figures 10–18, the vegetation is plotted in white. A preliminary simulation, under the same weather conditions, was performed on a wider domain (on the right of Figure 8), 2500 m long, 2500 m wide, and 250 m high (500 × 500 × 50-cell), which included the ocean on the south-east boundary, to obtain the wind profile shape to provide input to the current situation and planned situation simulations, in order to take into account the high buildings upstream the campus and, at the same time, reduce the computational time.

In previous studies, (Chiri and Giovagnorio 2012 [35]; Giovagnorio and Chiri, 2016 [36]) we verified the index of insulation even in the early stage of the design process. Nevertheless, in this case, due to both the dimension of the considered domains and the low height of the buildings, we decided to verify this parameter only at a smaller scale in future investigations. Data were firstly extracted from the output files of ENVI-met with the tool Leonardo, and then converted into 3D matrices for visualization and analysis with Matlab (Version R2020a, The MathWorks, Inc., Natick, MA, USA).

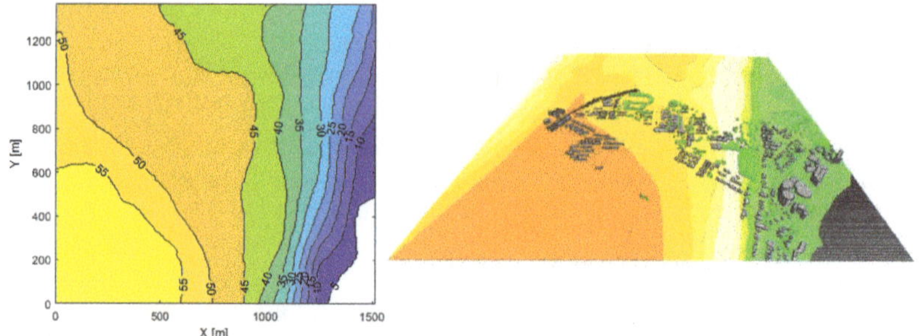

Figure 8. On the left: terrain elevation (m above the mean sea level) employed in the ENVI-met simulations; on the right: a view of the 3D model employed for the preliminary simulation on a larger scale.

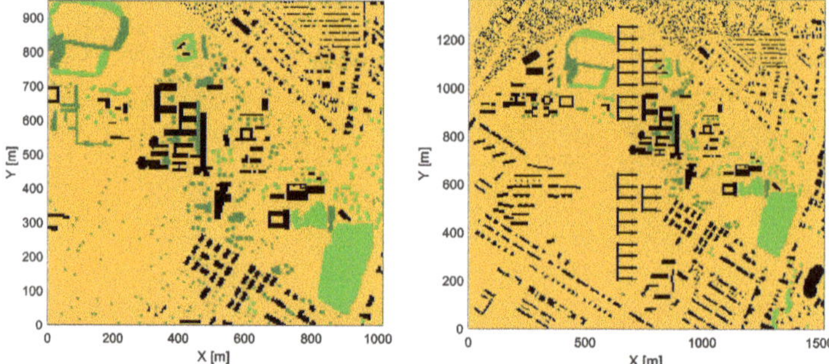

Figure 9. The domain employed in the ENVI-met simulations, with buildings (in black) and vegetation (pale green and dark green): on the left, the current situation (1020 m/204 cells in the X-direction, 920 m/191 cells in the Y-direction), on the right, the planned situation (1525 m/305 cells in the X-direction, 1375 m/275 cells in the Y-direction).

4. Results

4.1. Analysis and Comparison of Current Situation and Planned Situation

In Figures 10, 12, 14, 16 and 17, the maps of the simulated relative humidity, wind speed, air temperature, PMV and PMV with saturated values, for the current situation (on the left) and for the planned situation (on the right), are shown in false colours: colours close to dark blue mean low values, close to dark red high values; to avoid any confusion with the data, the buildings are in black, the vegetation in white. The quantities are measured 1.50 m from the ground, i.e., at pedestrian level. To simplify the quantitative evaluation of the microclimatic performances of the new buildings when compared to the existing buildings' ones, in Figures 11, 13, 15 and 18, a zoom is shown of the proposed buildings in the northern complex (on the left) and in the southern complex (on the right), with counter lines for each quantity. Moreover, on the left plot of Figures 11, 13, 15 and 18, the existing buildings are shown as well. As the new buildings are planned to be placed downstream of the existing Campus, here the target is not to improve the microclimatic performances of the existing Campus but to design new buildings that have better performances than the existing ones. For this reason, a comparison between the values (of humidity, wind velocity, air temperature and PMV) in the planned buildings and in the empty space they are planned to be placed is not shown.

A first critical issue of the current situation is the humidity (on the left of Figure 10), as the botanical garden (the large white stain on the south-east corner of the Figure) significantly influences the comfort performances of the Campus public space in terms of rise of humidity index. The humidity produced by plant transpiration increases the air humidity and the prevalent wind, coming from the sea in the south-east direction, it is transferred across the Campus, affecting all the downstream buildings and public spaces; this issue had to be taken into account in the planning of the distribution of the new buildings, as in a sub-tropical climate an increase in humidity can be very unpleasant. For this reason, the new buildings were split into two complexes, one south of and one north of the high humidity wake, in order to avoid an overlap with the humidity wake. Consequently, the new buildings have lower humidity values than the existing ones. As a matter of fact, the humidity values in the courtyards of the proposed southern complex (on the right of Figure 11) are between around 52.00% and 53.00%, in the proposed northern complex (on the left of Figure 11) between around 52.50% and 52.75%, while in the courtyards of the existing Campus (on the left of Figure 11) the values are almost everywhere higher than 53.00%, with peaks of 55.00%.

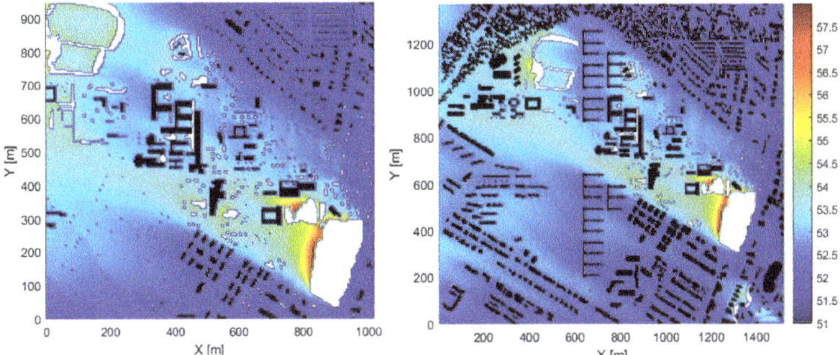

Figure 10. Relative humidity [%] at pedestrian level: on the left, in the current situation; on the right, in the planned situation.

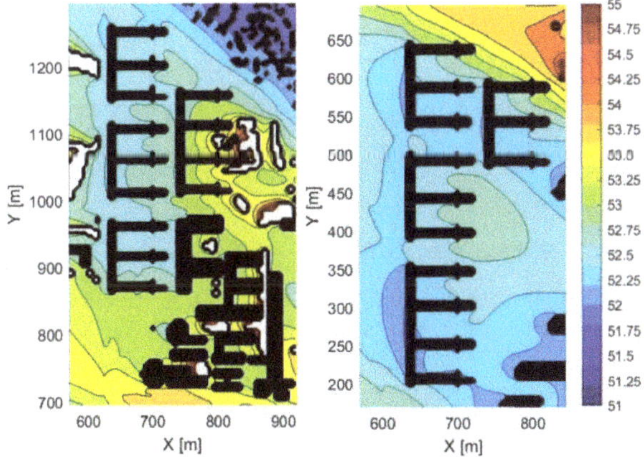

Figure 11. Comparison of the relative humidity [%] at pedestrian level in the new buildings and in the existing ones: on the left, the northern complex of proposed buildings with the old Campus; on the right, the southern complex of proposed buildings.

A second critical issue of the current situation is the ventilation in the courtyards of the existing Campus (plot on the left of Figure 12: colours represent the absolute values, arrows both the direction and the absolute values). The wind velocity is damped from the trees of the botanical garden and from the downstream buildings such that, even if the existing 'C'-shaped block are correctly east oriented, most of the courtyards experience velocities lower than around 1.5–2 m/s. The worst cases are in the courtyards downstream of the north-south oriented building on the right of the Campus, which, together with the trees present there, reduce the wind velocity to less than 1 m/s. To overcome this issue, two strategies have been pursued: the already-mentioned split into two complexes and the employment of larger courtyards to enhance the air circulation and, in turn, improve the courtyards' overall comfort. The plot on the right of Figure 12 confirms that this result has been achieved, as the wind velocities inside the new buildings courtyards are higher almost everywhere than in the existing ones. In particular, the contour plots in Figure 13 show that the wind velocity in the courtyards of the proposed southern complex (on the right of Figure 13) are between around 2.0 m/s and 4.0 m/s (somewhere higher than 4.0 m/s), in the proposed northern complex (on the left of Figure 13) mainly between around 1.5 m/s and 3.0 m/s, while in the courtyards of the existing Campus (on the left in Figure 13), the values are almost everywhere lower than around 1.5 m/s.

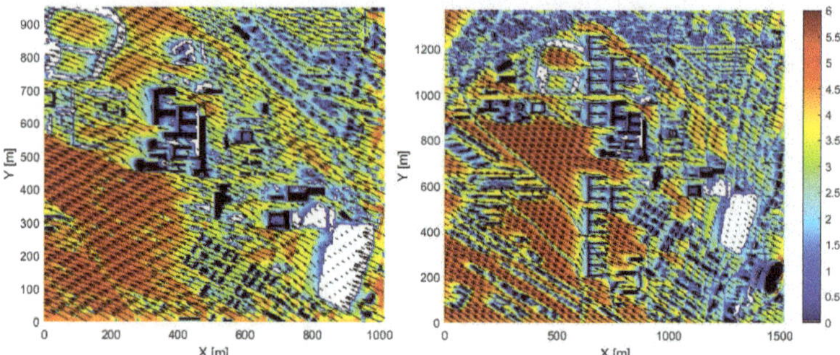

Figure 12. Wind velocity [m/s] at pedestrian level: on the left, in the current situation; on the right, in the planned situation.

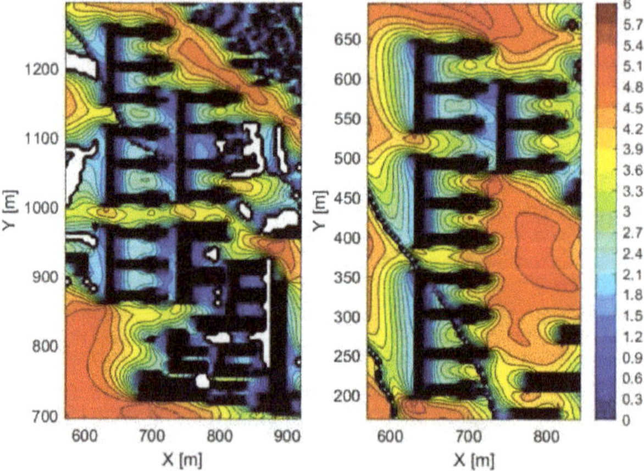

Figure 13. Comparison of wind velocity [m/s] at pedestrian level in the new buildings and in the existing ones: on the left, the northern complex of proposed buildings with the old Campus; on the right, the southern complex of proposed buildings.

The left of Figure 14 shows that temperature is not a relevant issue: the average temperature range is not very wide (around 24.0 °C in almost all the existing Campus area) and does not exceed typical tropical conditions. In the new buildings, temperature (on the right of Figure 14) does not get better, and it tends to remain on the average of tropical climate. Figure 15 shows that the air temperature is higher in the courtyards of the proposed southern complex (on the right of Figure 15), between around 24.15 °C and 24.00 °C, and slightly lower in the proposed northern complex between and in the existing Campus (on the left of Figure 15), around 24.00 °C and 23.95 °C.

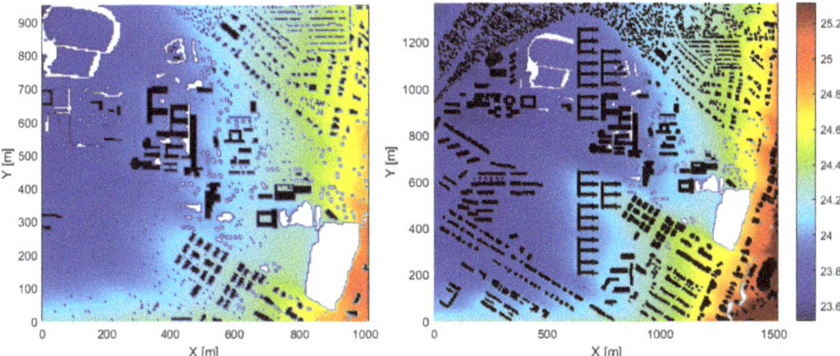

Figure 14. Air temperature [°C] at pedestrian level: on the left, in the current situation; on the right, in the planned situation.

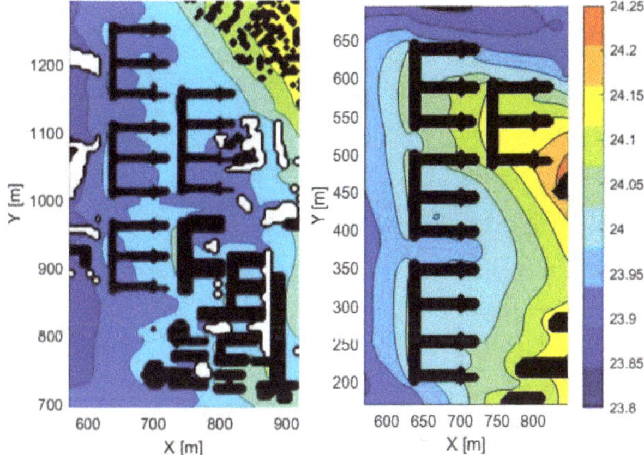

Figure 15. Comparison of the air temperature [°C] at pedestrian level in the new buildings and in the existing ones: on the left, the northern complex of proposed buildings with the old Campus; on the right, the southern complex of proposed buildings.

The value of the PMV index is very relevant in microclimate design, because, as mentioned above, PMV summarizes the external thermal comfort conditions in a single number. The left plot of Figure 16 highlights that, in the current situation, a non-optimal situation is found, as in the public spaces are almost everywhere around 2.0, meaning that users will tend to feel warm and to perceive moderate heat stress. The right-hand side of Figure 16 shows that bigger courtyards improve overall comfort: the PMV index is now better distributed compared with the current situation, and it is dramatically improved in some local situations (e.g., inside almost all the new courtyards). To better highlight this point, in Figure 17, the colour map is saturated at 1.55 and 2.00: in this way, it is possible to see how the PMV values in most of the new courtyards have now dropped to values lower than 1.55, and all of them demonstrate better outdoor comfort performances than the current situation ones. To be more precise, Figure 18 shows the contour plots of PMV: a user will feel more comfortable in the courtyards of the proposed southern complex (PMV drops under 1.50 in some points, corresponding to a thermal perception of slightly warm and to a slight heat stress) and of the proposed northern complex (with slightly higher values) than in the existing Campus ones (where almost everywhere PMV is 1.90). This is mainly due to two aspects: the better ventilation and the lower humidity of the 'C'-shaped block east oriented (in particular for the ones not downstream of the current buildings).

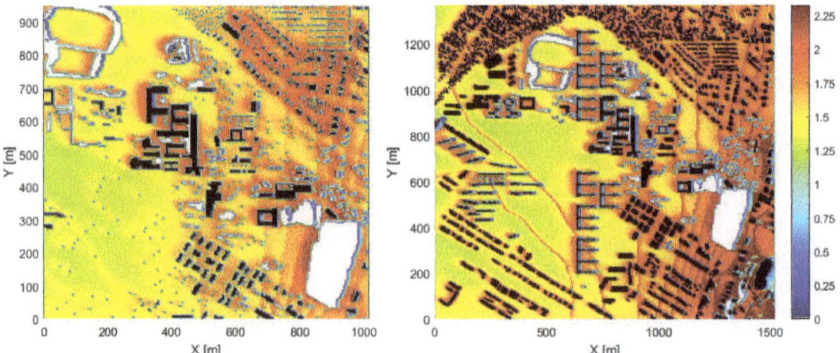

Figure 16. PMV [-]at pedestrian level: on the left, in the current situation; on the right, in the planned situation.

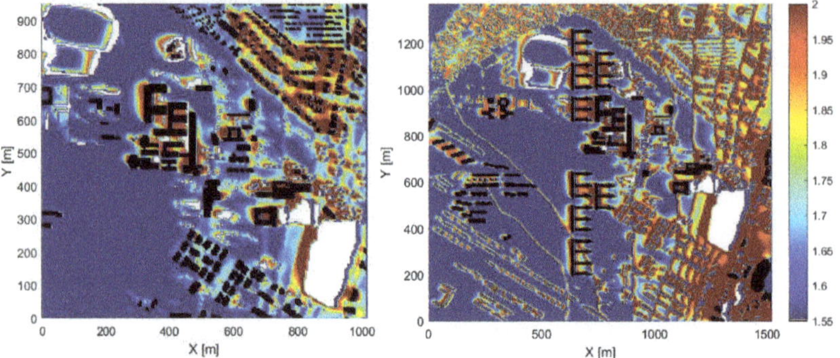

Figure 17. PMV (1.55 < PMV < 2.00) at pedestrian level: on the left, in the current situation; on the right, in the planned situation.

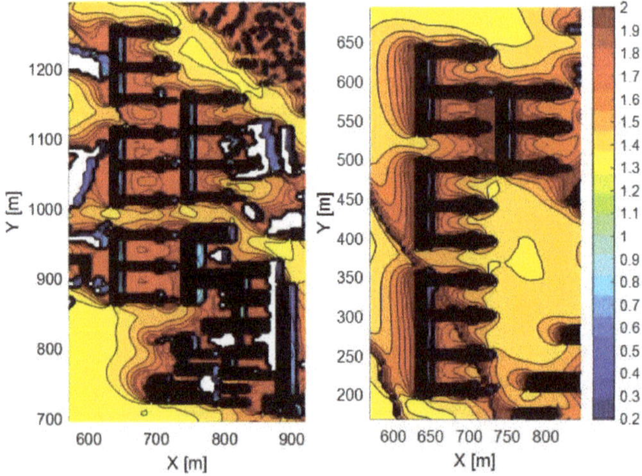

Figure 18. Comparison of the PMV [-] at pedestrian level in the new buildings and in the existing ones: on the left, the northern complex of proposed buildings with the old Campus; on the right, the southern complex of proposed buildings.

4.2. Process Advantages and Limitations

The positive results obtained following the analysis confirmed not only the validity of the project interventions, but the effectiveness of the design procedure. The integration of a continuous check system, placed side-by-side with the design process, assisted the designer in selecting and editing the typo-morphological solutions to allow definition of a more efficient urban form even in the early stages of the project. The advantages of this method allowed the technician to act upstream of the design process rather than downstream, to define more energy-saving and environmentally aware solutions. The software used reveals to be an extraordinary tool through which it is possible to qualitatively assess the trends stemming from the initial choices, directing the design towards the overall improvement of its performance. The accuracy of the results is still related to the size of the area, and is reduced in direct ratio to the extension of the surface. Specifically, the numerical precision of the result is reduced to a qualitative trend, especially on the edges, while in the centre the data have better accuracy. If, on the one hand, the procedure applied to the case study demonstrates still to be in an experimental stage, on the other hand, it certifies its effectiveness if applied from the preliminary stages of the project to understand design behaviour resulting from initial choices. To build according to principles of sustainability does not mean sacrificing quality of spaces. Architects and technicians do not necessarily give up their own responsibilities by delegating architectural issues to computer and informatics technology. In this process, the software assists the designer as a technical instrument, but it cannot mechanically replace him with respect to design choices aimed at quality and at the spatial configuration of city. However, the results, confirmed by the experiment as expected, demonstrate the method's validity and its applicability in diffuse forms.

5. Discussion. The Results of Planning Intervention

As mentioned in the previous paragraph, even if some important improvement were set in terms of typology and spatial organization, the 'comb-shaped' layout of the new proposal is largely consistent with the previous one. In terms of macro-behaviour, the climatic performances of the two proposals have a lot of similarities. In fact, the overall 'qualitative analysis' of climatic performances of the two evaluated configuration (the current one and the 'Project Pais' design) reports a lot of similarities, except for a clear prevalence for the newest one with regards with the air-flow in the courtyards. This enhancement is largely derived from the H/W ratio adopted for the new project, which is closer to that of the Forjaz block even if the general dimension of the court is bigger. In fact, the existing building has a H/W ratio of 2.09 due to the fact that it has the same courtyard as the Forjaz one, but with higher buildings. The planned block has an H/W ratio of 2.70, which is slightly more than the existing one, even if it is less than 3.36 H/W of the Forjaz's block. The shape of the urban blocks affects the settlement's ability to use natural lighting and ventilation for indoor and outdoor spaces and, moreover, the urban design orientation with respect to the prevailing wind direction flow is dramatically important. Consequently, the confirmation of Forzjaz's 'comb-shaped' design is correct, both for general spatial arguments and for climatic comfort consideration. The modification applied to the original block shows its validity in optimizing overall performances of the courtyards and also their liveability as public spaces. The results of this work contribute to confirm the validity of an integrated approach to urban design. Nevertheless, the project should be supported by a deeper analysis of the relation between the quality of outer spaces (in terms of vegetation, water pools, etc.) and the micro-climate. Finally, as mentioned in previous paragraphs, this work does not consider insulation because of the width of the dominion; nevertheless, it is crucial in a closer investigation. Furthermore, the characteristics of materials can play a crucial role in determining a good comfort performance and in reducing thermal control and surface reflection. In summary, the future perspectives of this work may be summarized as follows:

1. A more specific work on the shape of the buildings that compose the blocks in order to verify the opportunity to introduce some interruption in the blocks to provide better air flow;

2. A focus on design intermediate scale, which may highlight the relation between the shape and the solar radiation;
3. A sort of 'close-up' on a group of buildings, testing different façade designs in order to prevent bad insulation of inner space or, conversely, too much radiation on exposed walls.
4. A deeper test on the influence of vegetation and pools in outer spaces in microclimate performances.
5. Research on the influence of the albedo of local materials on the climatic building' behaviour.

6. Conclusions

The Masterplan for the Campus of the University Eduardo Mondlane in Maputo (Mozambique) has been employed as a case study to test a recursive and multidisciplinary methodology that integrates architectural design with fluid dynamics, in an approach where the shape and distribution of buildings are planned following both the optimization of urban microclimate and the principles of architectural design. To the best of the authors' knowledge, this is the first time that this kind of architectural urban microclimate design process has been applied in the social and cultural context of the African city of Maputo (Mozambique).

The proposed planned intervention envisages C-shaped blocks, east-oriented, in harmony with those of Forjaz's Masterplan, but optimized in shape, proportions and distribution according to the results of the microclimate simulations on the present situation. The software ENVI-met (version 4.4), widely adopted and validated in recent years, was employed for the microclimatic simulations and the PMV index was used to compare the outdoor comfort performance of the planned intervention to the ones of the present situation.

The results show that the planned C-shaped blocks, east-oriented and with higher H/W ratio and courtyard ratio than the existing buildings, allow better natural ventilation without increasing the temperature. The splitting of the Campus into a northern and a southern complex avoids the planned buildings being affected by the high-humidity wake generated when the prevailing wind passes through the botanical garden. PMV fields confirm the good performance, in term of outdoor comfort, of the planned buildings, as the PMV value in their courtyards is always better than the one in the existing Campus. As a consequence, we can state that the planned solution has better microclimatic performances, also in terms of outdoor comfort, than the existing one.

These results were reached thanks to the above-mentioned recursive and multidisciplinary process, where urban planning is not only guided by architectural decisions (with the risk to have an interesting but inefficient urban environments), but also, conversely, not only guided by the "mechanical" answers given by a software (with the risk to have an efficient but unpleasant urban environments).

Author Contributions: Conceptualization, G.M.C. and S.F.; methodology, G.M.C., S.F. and L.T.; software, S.F. and L.T.; formal analysis investigation, G.M.C., A.C. and L.N.; resources, L.N.; data curation, L.T.; writing—original draft preparation, G.M.C., S.F. and A.C.; writing—review and editing, M.A.; visualization, L.T. and L.N.; supervision, G.M.C.; project administration, M.A.; funding acquisition, M.A. and G.M.C. All authors have read and agreed to the published version of the manuscript.

Funding: This work is funded by the Autonomous Region of Sardinia within the program of international cooperation ex L.19/96- "Project Païs" 2019.

Acknowledgments: This project was developed by a joint group of Cagliari Faculty of Engineering and Architecture—Department of Civil and Environmental Engineering and Architecture and the University of Maputo Eduardo Mondlane, Faculdade de Arquitectura y Planeamento Fisico within the "Project País" financed in 2019 by the Autonomous Region of Sardinia within the program of international cooperation ex L.19/96. We would like to mention: Sara Spiga for her thesis project whose work largely contributed to the general Masterplan; Marta Pilleri who spent a lot of time in Maputo during her PhD in order to collect data; Vania Armando who enthusiastically contributed on public spaces design; all the colleagues and students of both Faculties for supporting us in the development of the project.

Conflicts of Interest: The authors declare no conflict of interest. The funders had no role in the design of the study; in the collection, analyses, or interpretation of data; in the writing of the manuscript, or in the decision to publish the results.

References

1. De Pascali, P. *Città ed Energia. La Valenza Energetica dell'Organizzazione Insediativa*; Franco Angeli: Milano, Italy, 2008.
2. Balaras, C.; Droutsa, K.G.; Dascalaki, E.; Kontoyiannidis, S.; Moro, A.; Bazzan, E. Urban Sustainability Audits and Ratings of the Built Environment. *Energies* **2019**, *12*, 4243. [CrossRef]
3. Susowake, Y.; Masrur, H.; Yabiku, T.; Senjyu, T.; Howlader, A.M.; Abdel-Akher, M.; Hemeida, A.M. A Multi-Objective Optimization Approach towards a Proposed Smart Apartment with Demand-Response in Japan. *Energies* **2019**, *13*, 127. [CrossRef]
4. Cordero, A.S.; Melgar, S.G.; Andújar, J.M. Green Building Rating Systems and the New Framework Level(s): A Critical Review of Sustainability Certification within Europe. *Energies* **2019**, *13*, 66. [CrossRef]
5. Mancini, F.; Basso, G.L. How Climate Change Affects the Building Energy Consumptions Due to Cooling, Heating, and Electricity Demands of Italian Residential Sector. *Energies* **2020**, *13*, 410. [CrossRef]
6. Santamouris, M. *Energy and Climate in the Urban Built Environment*; Informa UK Limited: London, UK, 2013.
7. Oke, T.R. The energetic basis of the urban heat island. *Q. J. R. Meteorol. Soc.* **1982**, *108*, 1–24. [CrossRef]
8. Chiri, G.; Giovagnorio, I. Gaetano Vinaccia's (1881–1971) Theoretical Work on the Relationship between Microclimate and Urban Design. *Sustainability* **2015**, *7*, 4448–4473. [CrossRef]
9. Giovagnorio, I.; Usai, D.; Palmas, A.; Chiri, G. The environmental elements of foundations in Roman cities: A theory of the architect Gaetano Vinaccia. *Sustain. Cities Soc.* **2017**, *32*, 42–55. [CrossRef]
10. Badas, M.G.; Ferrari, S.; Garau, M.; Seoni, A.; Querzoli, G. On the Flow Past an Array of Two-Dimensional Street Canyons Between Slender Buildings. *Boundary-Layer Meteorol.* **2020**, *174*, 251–273. [CrossRef]
11. Garau, M.; Badas, M.G.; Ferrari, S.; Seoni, A.; Querzoli, G. Air exchange in urban canyons with variable building width: A numerical LES approach. *Int. J. Environ. Pollut.* **2019**, *65*, 103. [CrossRef]
12. Badas, M.G.; Salvadori, L.; Garau, M.; Querzoli, G.; Ferrari, S. Urban areas parameterisation for CFD simulation and cities air quality analysis. *Int. J. Environ. Pollut.* **2019**, *66*, 14. [CrossRef]
13. Garau, M.; Badas, M.G.; Ferrari, S.; Seoni, A.; Querzoli, G. Turbulence and Air Exchange in a Two-Dimensional Urban Street Canyon Between Gable Roof Buildings. *Boundary-Layer Meteorol.* **2018**, *167*, 123–143. [CrossRef]
14. Badas, M.G.; Ferrari, S.; Garau, M.; Querzoli, G. On the effect of gable roof on natural ventilation in two-dimensional urban canyons. *J. Wind. Eng. Ind. Aerodyn.* **2017**, *162*, 24–34. [CrossRef]
15. Ferrari, S.; Badas, M.G.; Garau, M.; Seoni, A.; Querzoli, G. The air quality in narrow two-dimensional urban canyons with pitched and flat roof buildings. *Int. J. Environ. Pollut.* **2017**, *62*, 22. [CrossRef]
16. Costanzo, V.; Yao, R.; Xu, T.; Xiong, J.; Zhang, Q.; Li, B. Natural ventilation potential for residential buildings in a densely built-up and highly polluted environment. A case study. *Renew. Energy* **2019**, *138*, 340–353. [CrossRef]
17. Ng, E. (Ed.) *Designing High-Density Cities for Social and Environmental Sustainability*; Earthscan: London, UK, 2010.
18. Coccolo, S.; Kämpf, J.; Scartezzini, J.-L.; Pearlmutter, D. Outdoor human comfort and thermal stress: A comprehensive review on models and standards. *Urban Clim.* **2016**, *18*, 33–57. [CrossRef]
19. Fanger, P.O. *Thermal Comfort, Analysis and Applications in Environmental Engineering*; McGraw-Hill Book Company: New York, NY, USA, 1970.
20. Bruse, M.; Fleer, H. Simulating surface–plant–air interactions inside urban environments with a three dimensional numerical model. *Environ. Model. Softw.* **1998**, *13*, 373–384. [CrossRef]
21. ENVI-met Homepage. Available online: http://www.envi-met.com (accessed on 15 July 2011).
22. Abdi, B.; Hami, A.; Zarehaghi, D. Impact of small-scale tree planting patterns on outdoor cooling and thermal comfort. *Sustain. Cities Soc.* **2020**, *56*, 102085. [CrossRef]
23. Ouali, K.; El Harrouni, K.; Abidi, M.L.; Diab, Y. Analysis of Open Urban Design as a tool for pedestrian thermal comfort enhancement in Moroccan climate. *J. Build. Eng.* **2020**, *28*, 101042. [CrossRef]
24. Hassan, S.A.; Abrahem, S.A.; Husian, M.S. Comparative Analysis of Housing Cluster Formation on Outdoor Thermal Comfort in Hot-arid Climate. *J. Adv. Res. Mech. Therm. Sci.* **2019**, *63*, 10.
25. Limona, S.S.; Al-Hagla, K.S.; El-Sayad, Z.T. Using simulation methods to investigate the impact of urban form on human comfort. Case study: Coast of Baltim, North Coast, Egypt. *Alex. Eng. J.* **2019**, *58*, 273–282. [CrossRef]
26. Shinzato, P.; Simon, H.; Duarte, D.H.S.; Bruse, M. Calibration process and parametrization of tropical plants using ENVI-met V4—Sao Paulo case study. *Arch. Sci. Rev.* **2019**, *62*, 112–125. [CrossRef]

27. Karakounos, I.; Dimoudi, A.; Zoras, S. The influence of bioclimatic urban redevelopment on outdoor thermal comfort. *Energy Build.* **2018**, *158*, 1266–1274. [CrossRef]
28. Nasrollahi, N.; Hatami, M.; Khastar, S.R.; Taleghani, M. Numerical evaluation of thermal comfort in traditional courtyards to develop new microclimate design in a hot and dry climate. *Sustain. Cities Soc.* **2017**, *35*, 449–467. [CrossRef]
29. GhaffarianHoseini, A.; Berardi, U.; GhaffarianHoseini, A. Thermal performance characteristics of unshaded courtyards in hot and humid climates. *Build. Environ.* **2015**, *87*, 154–168. [CrossRef]
30. Fabbri, K.; Costanzo, V. Drone-assisted infrared thermography for calibration of outdoor microclimate simulation models. *Sustain. Cities Soc.* **2020**, *52*, 101855. [CrossRef]
31. Blocken, B. Computational Fluid Dynamics for urban physics: Importance, scales, possibilities, limitations and ten tips and tricks towards accurate and reliable simulations. *Build. Environ.* **2015**, *91*, 219–245. [CrossRef]
32. Ferreira, M.J.; Oliveira, A.; Soares, J.; Codato, G.; Barbaro, E.; Escobedo, J.F. Radiation balance at the surface in the city of São Paulo, Brazil: Diurnal and seasonal variations. *Theor. Appl. Clim.* **2011**, *107*, 229–246. [CrossRef]
33. Fabbri, K.; Canuti, G.; Ugolini, A. A methodology to evaluate outdoor microclimate of the archaeological site and vegetation role: A case study of the Roman Villa in Russi (Italy). *Sustain. Cities Soc.* **2017**, *35*, 107–133. [CrossRef]
34. Duarte, D.H.S.; Shinzato, P.; Gusson, C.D.S.; Alves, C.A. The impact of vegetation on urban microclimate to counterbalance built density in a subtropical changing climate. *Urban Clim.* **2015**, *14*, 224–239. [CrossRef]
35. Chiri, G.M.; Giovagnorio, I. The Role of the City's Shape in Urban Sustainability. *Int. Trans. J. Eng. Manag. Appl. Sci. Technol.* **2012**, *3*, 14.
36. Giovagnorio, I.; Chiri, G. The Environmental Dimension of Urban Design: A Point of View. In *Sustainable Urbanization*; IntechOpen: London, UK, 2016.

MDPI

St. Alban-Anlage 66

4052 Basel

Switzerland

Tel. +41 61 683 77 34

Fax +41 61 302 89 18

www.mdpi.com

Energies Editorial Office

E-mail: energies@mdpi.com

www.mdpi.com/journal/energies